Network Games

Theory, Models, and Dynamics

Synthesis Lectures on Communication Networks

Editor
Jean Walrand, *University of California, Berkeley*

Synthesis Lectures on Communication Networks is an ongoing series of 50- to 100-page publications on topics on the design, implementation, and management of communication networks. Each lecture is a self-contained presentation of one topic by a leading expert. The topics range from algorithms to hardware implementations and cover a broad spectrum of issues from security to multiple-access protocols. The series addresses technologies from sensor networks to reconfigurable optical networks. The series is designed to:

- Provide the best available presentations of important aspects of communication networks.

- Help engineers and advanced students keep up with recent developments in a rapidly evolving technology.

- Facilitate the development of courses in this field

Communication Networks: A Concise Introduction
Jean Walrand and Shyam Parekh
2010

Path Problems in Networks
John S. Baras and George Theodorakopoulos
2010

Performance Modeling, Loss Networks, and Statistical Multiplexing
Ravi R. Mazumdar
2009

Network Simulation
Richard M. Fujimoto, Kalyan S. Perumalla, and George F. Riley
2006

Network Games: Theory, Models, and Dynamics
Ishai Menache and Asuman Ozdaglar

ISBN: 978-3-031-79247-2   paperback
ISBN: 978-3-031-79248-9   ebook

DOI 10.1007/978-3-031-79248-9

A Publication in the Springer series
SYNTHESIS LECTURES ON COMMUNICATION NETWORKS

Lecture #9
Series Editor: Jean Walrand, *University of California, Berkeley*
Series ISSN
Synthesis Lectures on Communication Networks
Print 1935-4185   Electronic 1935-4193

Network Games

Theory, Models, and Dynamics

Ishai Menache
Microsoft Research New England

Asuman Ozdaglar
Massachusetts Institute of Technology

SYNTHESIS LECTURES ON COMMUNICATION NETWORKS #9

ABSTRACT

Traditional network optimization focuses on a single control objective in a network populated by obedient users and limited dispersion of information. However, most of today's networks are large-scale with lack of access to centralized information, consist of users with diverse requirements, and are subject to dynamic changes. These factors naturally motivate a new distributed control paradigm, where the network infrastructure is kept simple and the network control functions are delegated to individual agents which make their decisions independently ("selfishly"). The interaction of multiple independent decision-makers necessitates the use of game theory, including economic notions related to markets and incentives.

This monograph studies game theoretic models of resource allocation among selfish agents in networks. The first part of the monograph introduces fundamental game theoretic topics. Emphasis is given to the analysis of *dynamics* in game theoretic situations, which is crucial for design and control of networked systems. The second part of the monograph applies the game theoretic tools for the analysis of resource allocation in communication networks. We set up a general model of routing in wireline networks, emphasizing the congestion problems caused by delay and packet loss. In particular, we develop a systematic approach to characterizing the inefficiencies of network equilibria, and highlight the effect of autonomous service providers on network performance. We then turn to examining distributed power control in wireless networks. We show that the resulting Nash equilibria can be efficient if the degree of freedom given to end-users is properly designed.

KEYWORDS

game theory, Nash equilibrium, dynamics, communication networks, routing, power control

To Helena, Sophia, and Rami

Ishai Menache

To Daron and my parents for their
unconditional love and support

Asuman Ozdaglar

Contents

Preface

Traditional network optimization focuses on a well-defined control objective in a network populated by obedient users and limited dispersion of information, and uses convex optimization techniques to determine efficient allocation of resources (see [8], [20], [22], [102]). Most of today's networked systems, such as the Internet, transportation networks, and electricity markets, differ from this model in their structure and operation. First, these networks are large-scale with lack of access to centralized information and subject to dynamic changes. Hence, control policies have to be decentralized, scalable, and robust against unexpected disturbances. Second, these networks consist of interconnection of heterogeneous autonomous entities and serve users with diverse requirements, so there is no central party with enforcement power or accurate information about user needs. This implies that selfish incentives and private information of users need to be incorporated into the control paradigm. Finally, these networks are subject to continuous upgrades and investments in new technologies, making economic incentives of service and content providers much more paramount.

These new challenges have naturally motivated a new distributed control paradigm, where the network infrastructure is kept simple and the network control functions are delegated to individual agents, which make their decisions independently ("selfishly"), according to their own performance objectives. The key aspect of this approach is to view the network as a resource to be shared by a number of heterogeneous users with different service requirements. The interaction of multiple independent decision-makers necessitates the use of game theory (the study of multi-agent problems) and also some ideas from economics related to markets and incentives. Consequently, the recent engineering literature considers a variety of game-theoretic and economic market models for resource allocation in networks.

This monograph studies game theoretic models for analysis of resource allocation among heterogeneous agents in networks. Given the central role that game theory plays in the analysis of networked systems, the first part of the monograph will be devoted to a systematic analysis of fundamental game theoretic topics. Emphasis will be placed on game theoretic tools that are commonly used in the analysis of resource allocation in current networks or those that are likely to be used in future networks. We start in the next chapter with strategic form games, which constitute the foundation of game theoretic analysis and enable us to introduce the central concept of Nash equilibrium, which makes predictions about equilibrium behavior in situations in which several agents interact. We also introduce several related concepts, such as correlated equilibria, both to clarify the conditions under which Nash equilibria are likely to provide a good approximation to behavior in various different circumstances and as alternative concepts of equilibrium that might be useful in the communication area in future research.

The Nash equilibrium is by its very definition a static concept, and as such, the study of its properties does not cover the analysis of *dynamics*, namely, if and how an equilibrium is reached. Issues of dynamic resource allocation and changes in behavior of users in wireline and wireless networks are of central importance in the analysis of communication networks. We present in Chapter 2 two complementary ways of introducing dynamic analysis in game theoretic situations. First, we study extensive form (dynamic) games where the relevant concept of (subgame perfect) Nash equilibrium will exhibit some dynamics itself. Second, we look at dynamics induced by the repeated play of the same strategic form game when agents are not so sophisticated to play Nash equilibrium, but follow simple myopic rules or rules of thumb. The game theory literature has established that this type of myopic play has several interesting properties in many games and, in fact, converges to Nash equilibrium for certain classes of games. Interestingly, these classes of games, which include potential games and supermodular games, have widespread applications for network game models, so the last part of this chapter includes a detailed analysis of potential and supermodular games and the dynamics of play under various myopic and reactive rules.

The second part of the monograph applies game theoretic tools to the analysis of resource allocation in wireline and wireless communication networks. Chapter 3 focuses on wireline networks, with special attention drawn to the consequences of selfish routing in communication networks. We first show how the notions of Nash equilibrium and subgame perfect Nash equilibrium enable us to develop simple models of routing and resource allocation in wireline networks. We set up a general model of routing in wireline networks emphasizing the congestion problems caused by delay and packet loss. We show how a variety of different models of congestion can be modeled as a static strategic form game with the cost of congestion captured by latency functions. We first establish existence of equilibria and provide basic characterization results. We then turn to the question of efficiency of equilibria in wireline communication problems. It is well known since the work of Alfred Pigou that equilibria in such situations with congestion problems can involve significant externalities. We provide examples illustrating these inefficiencies, both demonstrating that inefficiencies could be significantly very large (unbounded) and the possibilities of paradoxical result such as the Braess' paradox. We then develop a systematic approach to characterizing the inefficiencies of these equilibria.

Classic models of wireline communication ignore the fact that autonomous service providers are active participants in the flows of communication and do so by either (i) redirecting traffic within their own networks to achieve minimum intradomain total latency or (ii) charging prices for maximizing their individual revenues. We then show how the presence of autonomous service providers can be incorporated into the general framework of wireline games. We demonstrate how existence of pure strategy and mixed strategy equilibria can be established in the presence of service-provider routing and pricing, and develop an alternative mathematical approach to quantifying inefficiency of equilibria in several networking domains. The interesting set of results is that in a variety of network topologies, the presence of prices ameliorates the potential inefficiencies that exist in wireline communication networks. However, we also show that in complex communication networks, even in the presence of prices and optimal intradomain routing decisions, inefficiencies could be substantial, partly because of the double marginalization problem in the theory of oligopoly.

Chapter 4 deals with wireless communication games. Perhaps the most lucid example for the consequences of selfish behavior in wireless networks is the case where a mobile captures a shared collision channel by continuously transmitting packets, hence effectively nullifying other users' throughput. We show that this undesired scenario can be avoided under a natural power-throughput tradeoff assumption, where each user minimizes its average transmission rate (which is proportional to power investment) subject to minimum-throughput demand. An important element in our models is the incorporation of fading effects, assuming that the channel quality of each mobile is time-varying and available to the user prior to the transmission decisions. Our equilibrium analysis reveals that there are at most two Nash equilibrium points where all users obtain their demands, with one strictly better than the other in terms of power investment for all users. Furthermore, we suggest a fully distributed mechanism that leads to the better equilibrium. The above model is then extended to include wireless platforms where mobiles are allowed to autonomously control their transmission power. We demonstrate the existence of a power-superior equilibrium point that can be reached through a simple distributed mechanism. On the negative side, however, we point to the possibility of Braess-like paradoxes, where the use of multiple power levels can diminish system capacity and also lead to larger per-user power consumption, compared to the case where only a single level is permitted.

We conclude the monograph in Chapter 5 by outlining high-level directions for future work in the area of network games, incorporating novel challenges not only in terms of game-theoretic modeling and analysis, but also with regard to the proper exploitation of the associated tools in current and future networking systems.

Large parts of Chapters 3 and 4 are based on our individual research in the area. We are naturally indebted to our coauthors whose advice, knowledge, and deep insight have been invaluable. Ishai Menache wishes to particularly thank Nahum Shimkin for his guidance and collaboration. Asuman Ozdaglar would like to thank Daron Acemoglu for his collaboration and support.

We are also grateful to Ermin Wei for making detailed comments on drafts of this monograph. Finally, we wish to acknowledge the research support of NSF grants CMMI-0545910 and SES-0729361, AFOSR grantsFA9550-09-1-0420 and R6756-G2,ARO grant 56549NS, the DARPA ITMANET program, and a Marie Curie International Fellowship within the 7th European Community Framework Programme.

Ishai Menache and Asuman Ozdaglar
March 2011

PART I

Game Theory Background

CHAPTER 1

Static Games and Solution Concepts

This chapter presents the fundamental notions and results in noncooperative game theory. In Section 1.1, we introduce the standard model for static strategic interactions, the strategic form game. In Section 1.2, we define various solution concepts associated with strategic form games, including the Nash equilibrium. We then proceed in Sections 1.3–1.4 to address the issues of existence and uniqueness of a Nash equilibrium.

1.1 STRATEGIC FORM GAMES

We first introduce strategic form games (also referred to as normal form games). A *strategic form* game is a model for a static game in which all players act simultaneously without knowledge of other players' actions.

Definition 1.1 (Strategic Form Game) A strategic form game is a triplet $\langle \mathcal{I}, (S_i)_{i \in \mathcal{I}}, (u_i)_{i \in \mathcal{I}} \rangle$ where

1. $\mathcal{I}$ is a finite set of players, $\mathcal{I} = \{1, \ldots, I\}$.

2. S_i is a non-empty set of available actions for player i.

3. $u_i : S \to \mathbb{R}$ is the payoff (utility) function of player i where $S = \prod_i S_i$.[1]

     For strategic form games, we will use the terms *action* and *(pure) strategy* interchangeably.[2] We denote by $s_i \in S_i$ an action for player i, and by $s_{-i} = [s_j]_{j \neq i}$ a vector of actions for all players *except* i. We refer to the tuple $(s_i, s_{-i}) \in S$ as an *action (strategy) profile*, or *outcome*. We also denote by $S_{-i} = \prod_{j \neq i} S_j$ the set of actions (strategies) of all players except i. Our convention throughout will be that each player i is interested in action profiles that "maximize" his utility function u_i.

---

[1] Here we implicitly assume that players have preferences over the outcomes and that these preferences can be captured by assigning a utility function over the outcomes. Note that not all preference relations can be captured by utility functions (see Chapters 1,3 and 6 of [60] for more on this issue).

[2] We will later use the term "strategy" more generally to refer to randomizations over actions, or contingency plans over actions in the context of dynamic games.

The next two examples illustrate strategic form games with finite and infinite strategy sets.

Example 1.2 Finite Strategy Sets A two-player game where the strategy set S_i of each player is finite can be represented in matrix form. We adopt the convention that the rows (columns) of the matrix represent the action set of player 1 (player 2). The cell indexed by row x and column y contains a pair, (a, b), where a is the payoff to player 1 and b is the payoff to player 2, i.e., $a = u_1(x, y)$ and $b = u_2(x, y)$. This class of games is sometimes referred to as *bimatrix games*. For example, consider the following game of "Matching Pennies."

	HEADS	TAILS
HEADS	$-1, 1$	$1, -1$
TAILS	$1, -1$	$-1, 1$

Matching Pennies.

This game represents "pure conflict," in the sense that one player's utility is the negative of the utility of the other player, i.e., the sum of the utilities for both players at each outcome is "zero." This class of games is referred to as *zero-sum games* (or *constant-sum games*) and has been extensively studied in the game theory literature [15].

Example 1.3 Infinite Strategy Sets The strategy sets of players can also have infinitely many elements. Consider the following game of *Cournot Competition*, which models two firms producing the same homogeneous good and seeking to maximize their profits. The formal game $G = \langle \mathcal{I}, (S_i), (u_i) \rangle$ consists of:

1. A set of two players, $\mathcal{I} = 1, 2$.

2. A strategy set $S_i = [0, \infty)$ for each player i, where $s_i \in S_i$ represents the amount of good that the player produces.

3. A utility function u_i for each player i given by its total revenue minus its total cost, i.e.,

$$u_i(s_1, s_2) = s_i p(s_1 + s_2) - c_i s_i$$

where $p(q)$ represents the price of the good (as a function of the total amount of good q), and c_i is the unit cost for firm i.

For simplicity, we consider the case where both firms have unit cost, $c_1 = c_2 = 1$, and the price function is piecewise linear and is given by $p(q) = \max\{0, 2 - q\}$.

We analyze this game by considering the *best-response correspondences* for each of the firms. For firm i, the best-response correspondence $B_i(s_{-i})$ is a mapping from the set S_{-i} into set S_i such that

$$B_i(s_{-i}) = \{s_i \in S_i \mid u_i(s_i, s_{-i}) \geq u_i(s_i', s_{-i}), \text{ for all } s_i' \in S_i\}.$$

Note that for this example, the best response correspondences are unique-valued (hence can be referred to as best-response functions), since for all s_{-i}, there is a unique action that maximizes the utility function for firm i. In particular, we have:

$$B_i(s_{-i}) = \arg\max_{s_i \geq 0} (s_i\, p(s_i + s_{-i}) - s_i) .$$

When $s_{-i} > 1$, the action that maximizes the utility function of firm i, i.e., its best response, is 0. More generally, it can be seen (by using *first order optimality conditions*, see Appendix 1.B) that for any $s_{-i} \in S_{-i}$, the best response of firm i is given by

$$B_i(s_{-i}) = \arg\max_{s_i \geq 0} (s_i(2 - s_i - s_{-i}) - s_i)$$

$$= \begin{cases} \frac{1-s_{-i}}{2} & \text{if } s_{-i} \leq 1, \\ 0 & \text{otherwise.} \end{cases}$$

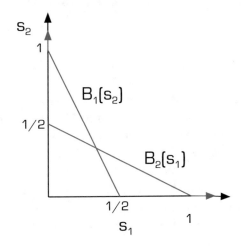

Figure 1.1: Best response functions for the Cournot Competition game.

Figure 1.1 illustrates the best response functions as a function of s_1 and s_2. Intuitively, we expect the outcome of this game to be at the point where both of these functions intersect. In the next section, we will argue that this intersection point is a reasonable outcome in this game.

1.2   SOLUTION CONCEPTS

This section presents the main solution concepts for strategic form games: strictly and weakly dominant and dominated strategies, pure and mixed Nash equilibrium, and correlated equilibrium.

1.2.1   DOMINANT AND DOMINATED STRATEGIES

In some games, it may be possible to predict the outcome assuming that all players are rational and fully knowledgeable about the structure of the game and each other's rationality. This is the case, for instance, for the well-studied Prisoner's Dilemma game. The underlying story of this game is as follows: Two people are arrested for a crime, placed in separate rooms, and questioned by authorities trying to extract a confession. If they both remain silent (i.e., cooperate with each other), then the authorities will not be able to prove charges against them and they will both serve a short prison term, say 2 years, for minor offenses. If only one of them confesses (i.e., does not cooperate), his term will be reduced to 1 year and he will be used as a witness against the other person, who will get a sentence of 5 years. If they both confess, they both get a smaller sentence of 4 years. This game can be represented in matrix form as follows:

	COOPERATE	DON'T COOPERATE
COOPERATE	$-2, -2$	$-5, -1$
DON'T COOPERATE	$-1, -5$	$-4, -4$

Prisoner's Dilemma.

In this game, regardless of the other players decision, playing DON'T COOPERATE yields a higher payoff for each player. Hence, the strategy DON'T COOPERATE is *strictly dominant*, i.e., no matter what strategy the other player chooses, this strategy always yields a strictly better outcome. We can therefore infer that both players will choose DON'T COOPERATE and spend the next four years in jail while if they both chose the strategy COOPERATE, they could have ended up in jail only for two years! Prisoner's Dilemma is a paradigmatic example of self-interested rational behavior not leading to jointly (socially) optimal outcomes. We will see in Chapters 3 and 4 that many network games exhibit such inefficiencies in equilibrium due to selfish nature of players.

A compelling notion of equilibrium in games would be the *dominant strategy equilibrium*, where each player plays a dominant strategy as formalized in the next definition.

Definition 1.4   Dominant Strategy   A strategy $s_i \in S_i$ is a *dominant strategy* for player i if

$$u_i(s_i, s_{-i}) \geq u_i(s_i', s_{-i}) \qquad \text{for all } s_i' \in S_i \text{ and for all } s_{-i} \in S_{-i}.$$

It is *strictly dominant* if this relation holds with a strict inequality.

Definition 1.5   Dominant Strategy Equilibrium   A strategy profile s^* is a *(strictly) dominant strategy equilibrium* if for each player i, s_i^* is a (strictly) dominant strategy.

In the Prisoner's Dilemma game, (DON'T COOPERATE, DON'T COOPERATE) is a strictly dominant strategy equilibrium. Though compelling, dominant strategy equilibria do not always exist, as illustrated by the Matching Pennies game (cf. Example 1.2) and the next example. Consider a slightly modified Prisoner's Dilemma game in which players also have the strategy SUICIDE leading to the following payoff structure:

	COOPERATE	DON'T COOPERATE	SUICIDE
COOPERATE	$-2, -2$	$-5, -1$	$0, -20$
DON'T COOPERATE	$-1, -5$	$-4, -4$	$-4, -20$
SUICIDE	$-20, 0$	$-20, -4$	$-20, -20$

Prisoner's Dilemma with Suicide.

This payoff matrix models a scenario in which if one player chooses the strategy SUICIDE, then, due to lack of witnesses, the other player gets off free if he remains silent (cooperates). In this game, there is no dominant strategy equilibrium because of the additional strategy SUICIDE. Notice, however, that the strategy SUICIDE is the *worst* possible option for a player, no matter what the other player does. In this sense, SUICIDE is strictly dominated by the other two strategies. More generally, we say that a strategy is *strictly dominated* for a player if there exists some other strategy that yields a strictly higher payoff regardless of the strategies of the other players.

Definition 1.6 Strictly Dominated Strategy   A strategy $s_i \in S_i$ is *strictly dominated* for player i if there exists some $s_i' \in S_i$ such that

$$u_i(s_i', s_{-i}) > u_i(s_i, s_{-i}) \qquad \text{for all } s_{-i} \in S_{-i} .$$

Next, we define a weaker version of dominated strategies.

Definition 1.7 Weakly Dominated Strategy   A strategy $s_i \in S_i$ is *weakly dominated* for player i if there exists some $s_i' \in S_i$ such that

$$u_i(s_i', s_{-i}) \geq u_i(s_i, s_{-i}) \qquad \text{for all } s_{-i} \in S_{-i} ,$$

and

$$u_i(s_i', s_{-i}) > u_i(s_i, s_{-i}) \qquad \text{for some } s_{-i} \in S_{-i} .$$

It is plausible to assume that no player chooses a strictly dominated strategy. Moreover, common knowledge of payoffs and rationality leads players to do iterated elimination of strictly dominated strategies, as illustrated next.

1.2.2   ITERATED ELIMINATION OF STRICTLY DOMINATED STRATEGIES

In the Prisoner's Dilemma with Suicide game, the strategy SUICIDE is a strictly dominated strategy for both players. Therefore, no rational player would choose SUICIDE. Moreover, if player 1 is certain that player 2 is rational, then he can eliminate her opponent's SUICIDE strategy. We can use a similar reasoning for player 2. After one round of elimination of strictly dominated strategies, we are back to the Prisoner's Dilemma game, which has a dominant strategy equilibrium. Thus, iterated elimination of strictly dominated strategies leads to a unique outcome, (DON'T COOPERATE, DON'T COOPERATE) in this modified game. We say that a game is *dominance solvable* if iterated elimination of strictly dominated strategies yields a unique outcome.

Consider next another game.

	LEFT	MIDDLE	RIGHT
UP	4, 3	5, 1	6, 2
MIDDLE	2, 1	8, 4	3, 6
DOWN	3, 0	9, 6	2, 8

Example for iterated elimination of strictly dominated strategies.

In this game, there are no strategies that are strictly dominated for player 1 (the row player). On the other hand, the strategy MIDDLE is strictly dominated by the strategy RIGHT for player 2 (the column player). Thus, we conclude that it is not rational for player 2 to play MIDDLE and we can therefore remove this column from the game, resulting in the following reduced game.

	LEFT	RIGHT
UP	4, 3	6, 2
MIDDLE	2, 1	3, 6
DOWN	3, 0	2, 8

Game after one removal of strictly dominated strategies.

Now, note that both strategies MIDDLE and DOWN are strictly dominated by the strategy UP for player 1, which means that both of these rows can be removed, resulting in the following game.

	LEFT	RIGHT
UP	4, 3	6, 2

Game after three iterated removals of strictly dominated strategies.

We are left with a game where player 1 does not have any choice in his strategies, while player 2 can choose between LEFT and RIGHT. Since LEFT will maximize the utility of player 2, we conclude that the only rational strategy profile in the game is (UP,LEFT).[3]

More formally, we define the procedure of *iterated elimination of strictly dominated strategies (or iterated strict dominance)* as follows:

- **Step 0:** For each i, let $S_i^0 = S_i$.

- **Step 1:** For each i, define

$$S_i^1 = \{s_i \in S_i^0 \quad | \quad \text{there does not exist } s_i' \in S_i^0 \text{ such that } u_i\left(s_i', s_{-i}\right) > u_i\left(s_i, s_{-i}\right) \text{ for all } s_{-i} \in S_{-i}^0\}.$$

  ...

- **Step k:** For each i, define

$$S_i^k = \{s_i \in S_i^{k-1} \quad | \quad \text{there does not exist } s_i' \in S_i^{k-1} \text{ such that } u_i\left(s_i', s_{-i}\right) > u_i\left(s_i, s_{-i}\right) \text{ for all } s_{-i} \in S_{-i}^{k-1}\}.$$

- For each i, define

$$S_i^\infty = \cap_{k=0}^{\infty} S_i^k.$$

It is immediate that this procedure yields a nonempty set of strategy profiles under some assumptions stated in the next theorem.

Theorem 1.8 *Suppose that either (1) each S_i is finite, or (2) each $u_i\left(s_i, s_{-i}\right)$ is continuous and each S_i is compact. Then S_i^∞ is nonempty for each i.*

We next apply iterated elimination of strictly dominated strategies to the Cournot Competition game (cf. Example 1.3). Recall that we use the notation S_i^k to denote the set of strategies of player i that survive iterated elimination of strictly dominated strategies at step k. In the first step, we note that both firms must choose a quantity between $[0, \infty)$, i.e.,

$$S_1^1 = [0, \infty),$$
$$S_2^1 = [0, \infty).$$

Since the range of the best response function of player 1 is $[0, 1/2]$, any strategy outside this range is never a best response and therefore is strictly dominated. The same reasoning holds for player 2. Thus, at the second step, we have

$$S_1^2 = [0, 1/2],$$
$$S_2^2 = [0, 1/2].$$

---

[3]One might worry that different orders for the removal of dominated strategies can yield different results. However, it can be shown that the order in which strategies are eliminated does not affect the set of strategies that survive iterated elimination of strictly dominated strategies.

Given that player 2 only chooses actions in the interval [0, 1/2], then player 1 can *restrict* the domain of his best response function to only these values. Using his best response function, this implies that the strategies outside the interval [1/4, 1/2] are strictly dominated for player 1. Applying the same reasoning for player 2, we obtain

$$S_1^3 = [1/4, 1/2],$$
$$S_2^3 = [1/4, 1/2].$$

(see Figure 1.2). It can be shown that in the limit, the endpoints of the intervals converge to the point where the two best response functions intersect. Hence, the Cournot Competition game is another example of a dominance solvable game. Most games, however, are not solvable by iterated strict dominance; therefore, we need a stronger equilibrium notion to predict the outcome.

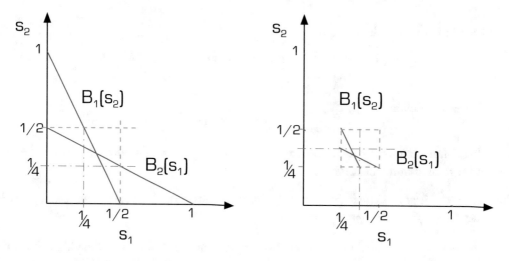

Figure 1.2: Elimination of strictly dominated strategies for the Cournot competition game.

1.2.3   NASH EQUILIBRIUM

We next introduce the fundamental solution concept for strategic form games, the notion of a *Nash equilibrium*. A Nash equilibrium captures a steady state of the play in a strategic form game such that each player acts optimally and forms correct conjectures about the behavior of the other players.

Definition 1.9   Nash Equilibrium   A *(pure strategy) Nash equilibrium* of a strategic form game $\langle \mathcal{I}, (S_i), (u_i)_{i \in \mathcal{I}} \rangle$ is a strategy profile $s^* \in S$ such that for all $i \in \mathcal{I}$, we have

$$u_i(s_i^*, s_{-i}^*) \geq u_i(s_i, s_{-i}^*) \qquad \text{for all } s_i \in S_i \,.$$

Hence, a Nash equilibrium is a strategy profile s^* such that no player i can profit by unilaterally deviating from his strategy s_i^*, assuming every other player j follows his strategy s_j^*. The definition of a Nash equilibrium can be restated in terms of the best-response correspondences.

Definition 1.10   Nash Equilibrium - Restated   Let $\langle \mathcal{I}, (S_i), (u_i)_{i \in \mathcal{I}} \rangle$ be a strategic game. For any $s_{-i} \in S_{-i}$, consider the best-response correspondence of player i, $B_i(s_{-i})$, given by

$$B_i(s_{-i}) = \{s_i \in S_i \mid u_i(s_i, s_{-i}) \geq u_i(s_i', s_{-i}) \text{ for all } s_i' \in S_i\} \,.$$

We say that an action profile s^* is a *Nash equilibrium* if

$$s_i^* \in B_i(s_{-i}^*) \qquad \text{for all } i \in \mathcal{I} \,.$$

This implies that for two player games, the set of Nash equilibria is given by the intersection of the best response correspondences of the two players (e.g., recall the Cournot Competition game). Below we give two other examples of games with pure strategy Nash equilibria.

Example 1.11   Battle of the Sexes
Consider a two player game with the following payoff structure:

	BALLET	SOCCER
BALLET	2, 1	0, 0
SOCCER	0, 0	1, 2

Battle of the Sexes.

This game, referred to as the Battle of the Sexes game, represents a scenario in which the two players wish to coordinate their actions, but have different preferences over their actions. This game has two pure Nash equilibria, i.e., the strategy profiles (BALLET, BALLET) and (SOCCER, SOCCER).

Example 1.12   Second Price Auction – with Complete Information
We consider a second price auction: There is a single indivisible object to be assigned to one of n players. Player i's valuation of the object is denoted by v_i. We assume without loss of generality that $v_1 \geq v_2 \geq \cdots \geq v_n > 0$ and that each player knows all the valuations $v_1, \ldots, v_n$, i.e., it is a complete information game.[4] The rules of this auction mechanism are described as follows:

- The players simultaneously submit bids, $b_1, .., b_n$.

---

[4]The analysis of the incomplete information version of this game, in which the valuations of other players are unknown (or probabilistically known), is similar.

- The object is given to the player with the highest bid (or to a random player among the ones bidding the highest value).

- The winner pays the *second* highest bid.

This mechanism induces a game among the players in which the strategy of each player is given by her bid, and her utility for a bid profile is given by her valuation of the object minus the price she pays, i.e., if player i is the winner, her utility is $v_i - b_j$ where j is the player with the second highest bid; otherwise, her utility is zero.

We first show that the strategy profile $(b_1, .., b_n) = (v_1, .., v_n)$ is a Nash equilibrium. First note that if indeed everyone plays according to this strategy profile, then player 1 receives the object and pays a price v_2. Hence, her payoff will be $v_1 - v_2 > 0$, and all other payoffs will be 0. Now, player 1 has no incentive to deviate since her utility cannot increase. Similarly, for all other players $i \neq 1$, in order for player i to change her payoff, she needs to bid more than v_1, in which case her payoff will be $v_i - v_1 < 0$. Therefore, no player has an incentive to unilaterally deviate, showing that this strategy profile is a Nash equilibrium.

We next show that the strategy profile $(v_1, 0, 0, ..., 0)$ is also a Nash equilibrium. As before, player 1 will receive the object, and will have a payoff of $v_1 - 0 = v_1$. Using the same argument as before, we conclude that none of the players have an incentive to deviate, and this strategy profile is a Nash equilibrium. We leave this as an exercise to show that the strategy profile $(v_2, v_1, 0, 0, ..., 0)$ is also a Nash equilibrium.

So far, we have shown that the game induced by the second price auction has multiple Nash equilibria. We finally show that for each player i, the strategy of bidding her valuation, i.e., $b_i = v_i$, in fact, weakly dominates all other strategies. Given a bid profile, let B^* denote the maximum of all bids excluding player i's bid, i.e.,

$$B^* = \max_{j \neq i} b_j.$$

Assume that player i's valuation is given by $v*$. Figure 1.3 illustrates the utility of player i as a function of B^*, when she bids her valuation, $b_i = v^*$, less than her valuation, $b_i < v^*$, and more than her valuation $b_i > v^*$. In the second graph, which represents the case when she bids $b_i < v^*$, notice that whenever $b_i \leq B^* \leq v^*$, player i receives zero utility since she loses the auction to whoever bid B^*. If she would have bid her valuation, she would have positive utility in this region (as depicted in the first graph). These figures show that for each player, bidding her own valuation weakly dominates all her other strategies.

An immediate implication of the preceding analysis is that there exist Nash equilibria (e.g., the strategy profile $(v_1, 0, 0, ..., 0)$) that involve the play of weakly dominated strategies.

Given that Nash equilibrium is a widely used solution concept in strategic form games, a natural question is why one should expect the Nash equilibrium outcome in a strategic form game. One justification is that since it represents a steady state situation, rational players should somehow reason their way to Nash equilibrium strategies; that is, Nash equilibrium might arise through

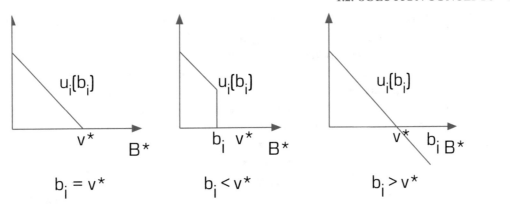

Figure 1.3: The utility of player i as a function of B^*, the maximum of all bids except i's bid.

introspection. This justification requires that players are rational and know the payoff functions of all players, that they know their opponents are rational and know the payoff functions, and so on. A second justification is that Nash equilibria are self-enforcing. That is, if players agree on a strategy profile before independently choosing their actions, then no player has an incentive to deviate if the agreed strategy profile is a Nash equilibrium. A final justification of the Nash equilibrium outcome is through learning dynamics or evolution, which will be discussed in Chapter 2.

1.2.3.1 Mixed Strategy and Mixed Strategy Nash Equilibrium
Recall the game of Matching Pennies:

	Heads	Tails
Heads	$1, -1$	$-1, 1$
Tails	$-1, 1$	$1, -1$

Matching Pennies.

It is easy to see that this game does not have a pure Nash equilibrium, i.e., for every pure strategy in this game, one of the parties has an incentive to deviate. However, if we allow the players to *randomize* over their choice of actions, we can determine a steady state of the play. Assume that player 1 picks Heads with probability p and Tails with probability $1 - p$, and that player 2 picks both Head and Tail with probability $\frac{1}{2}$. Then, with probability

$$\frac{1}{2}p + \frac{1}{2}(1 - p) = \frac{1}{2}$$

player 1 will receive a payoff 1. Similarly, she will receive a payoff -1 with the same probability. This implies that if player 2 plays the strategy $(\frac{1}{2}, \frac{1}{2})$, then no matter what strategy player 1 chooses, she will get the same payoff. Due to the symmetry of the game, we conclude that the randomized strategy $((\frac{1}{2}, \frac{1}{2}), (\frac{1}{2}, \frac{1}{2}))$ is a stable play of the game.

We next formalize the notion of "randomized" or mixed strategies. We first introduce some notation. Let Σ_i denote the set of probability measures over the pure strategy (action) set S_i. We use $\sigma_i \in \Sigma_i$ to denote the *mixed strategy* of player i. When S_i is a finite set, a mixed strategy is a finite dimensional probability vector, i.e., a vector whose elements denote the probability with which a particular action will be played. If S_i has two elements, the set of mixed strategies Σ_i is the one-dimensional probability simplex, i.e., $\Sigma_i = \{(x_1, x_2) \mid x_i \geq 0, \ x_1 + x_2 = 1\}$. We use $\sigma \in \Sigma = \prod_{i \in \mathcal{I}} \Sigma_i$ to denote a *mixed strategy profile*. Note that this implicitly assumes that players randomize independently. We similarly denote $\sigma_{-i} \in \Sigma_{-i} = \prod_{j \neq i} \Sigma_j$.

Following von Neumann-Morgenstern expected utility theory, we extend the payoff functions u_i from S to Σ by

$$u_i(\sigma) = \int_S u_i(s) d\sigma(s),$$

i.e., the payoff of a mixed strategy σ is given by the expected value of pure strategy payoffs under the distribution σ.

We are now ready to define the notion of a mixed strategy Nash equilibrium.

Definition 1.13 Mixed Strategy Nash Equilibrium   A mixed strategy profile σ^* is a *mixed strategy Nash equilibrium* (or mixed Nash equilibrium) if for each player i,

$$u_i(\sigma_i^*, \sigma_{-i}^*) \geq u_i(\sigma_i, \sigma_{-i}^*) \qquad \text{for all } \sigma_i \in \Sigma_i .$$

Note that since $u_i(\sigma_i, \sigma_{-i}^*) = \int_{S_i} u_i(s_i, \sigma_{-i}^*) d\sigma_i(s_i)$, it is sufficient to check only *pure* strategy "deviations" when determining whether a given profile is a Nash equilibrium. This leads to the following characterization of a mixed Nash equilibrium.

Proposition 1.14   *A mixed strategy profile σ^* is a (mixed strategy) Nash equilibrium if and only if for each player i,*

$$u_i(\sigma_i^*, \sigma_{-i}^*) \geq u_i(s_i, \sigma_{-i}^*) \qquad \text{for all } s_i \in S_i .$$

We also have the following useful characterization of a mixed Nash equilibrium in finite games (i.e., games with finite strategy sets).

Proposition 1.15   *Let $G = \langle \mathcal{I}, (S_i)_{i \in \mathcal{I}}, (u_i)_{i \in \mathcal{I}} \rangle$ be a finite strategic form game. Then, $\sigma^* \in \Sigma$ is a Nash equilibrium if and only if for each player $i \in \mathcal{I}$, every pure strategy in the support of σ_i^* is a best response to σ_{-i}^*.*

Proof. Let σ^* be a mixed strategy Nash equilibrium, and let $E_i^* = u_i(\sigma_i^*, \sigma_{-i}^*)$ denote the expected utility for player i. By Proposition 1.14, we have

$$E_i^* \geq u_i(s_i, \sigma_{-i}^*) \qquad \text{for all } s_i \in S_i.$$

We first show that $E_i^* = u_i(s_i, \sigma_{-i}^*)$ for all s_i in the support of σ_i^* (combined with the preceding relation, this proves one implication). Assume to arrive at a contradiction that this is not the case, i.e., there exists an action s_i' in the support of σ_i^* such that $u_i(s_i', \sigma_{-i}^*) < E_i^*$. Since $u_i(s_i, \sigma_{-i}^*) \leq E_i^*$ for all $s_i \in S_i$, this implies that

$$\sum_{s_i \in S_i} \sigma_i^*(s_i) u_i(s_i, \sigma_{-i}^*) < E_i^* \,,$$

– a contradiction. The proof of the other implication is similar and is therefore omitted. $\qquad\square$

It follows from this characterization that every action in the support of any player's equilibrium mixed strategy yields the same payoff. Note that this characterization result extends to games with infinite strategy sets: $\sigma^* \in \Sigma$ is a Nash equilibrium if and only if for each player $i \in \mathcal{I}$, given σ_{-i}^*, no action in S_i yields a payoff that exceeds his equilibrium payoff, and the set of actions that yields a payoff less than his equilibrium payoff has σ_i^*-measure zero.

Example 1.16 Let us return to the Battle of the Sexes game.

	BALLET	SOCCER
BALLET	2, 1	0, 0
SOCCER	0, 0	1, 2

Battle of the Sexes.

Recall that this game has 2 pure Nash equilibria. Using the characterization result in Proposition 1.15, we show that it has a *unique* mixed strategy Nash equilibrium (which is not a pure strategy Nash equilibrium). First, by using Proposition 1.15 (and inspecting the payoffs), it can be seen that there are no Nash equilibria where only one of the players randomizes over its actions. Now, assume instead that player 1 chooses the action BALLET with probability $p \in (0, 1)$ and SOCCER with probability $1 - p$, and that player 2 chooses BALLET with probability $q \in (0, 1)$ and SOCCER with probability $1 - q$. Using Proposition 1.15 on player 1's payoffs, we have the following relation

$$2 \times q + 0 \times (1 - q) = 0 \times q + 1 \times (1 - q) \,.$$

Similarly, we have

$$1 \times p + 0 \times (1 - p) = 0 \times p + 2 \times (1 - p) \,.$$

We conclude that the only possible mixed strategy Nash equilibrium is given by $q = \frac{1}{3}$ and $p = \frac{2}{3}$.

We conclude this section by discussing strict dominance by a mixed strategy. Consider the following game:

	LEFT	RIGHT
UP	2, 0	−1, 0
MIDDLE	0, 0	0, 0
DOWN	−1, 0	2, 0

Note that Player 1 has no pure strategy that strictly dominates MIDDLE. However, the mixed strategy $(\frac{1}{2}, 0, \frac{1}{2})$ of player 1 yields a strictly higher payoff than the pure strategy MIDDLE. In this case, we say that the strategy MIDDLE is strictly dominated by the strategy $(\frac{1}{2}, 0, \frac{1}{2})$.

Definition 1.17   Strict Domination by Mixed Strategies   A strategy s_i is *strictly dominated* for player i if there exists a mixed strategy $\sigma_i' \in \Sigma_i$ such that $u_i(\sigma_i', s_{-i}) > u_i(s_i, s_{-i})$, for all $s_{-i} \in S_{-i}$.

It can be shown that strictly dominated strategies are never used with positive probability in a mixed strategy Nash Equilibrium. In contrast, as we have seen in the Second Price Auction example, weakly dominated strategies can be used in a Nash Equilibrium.

1.2.4   CORRELATED EQUILIBRIUM

In a Nash equilibrium, players choose strategies (or randomize over strategies) independently. For games with multiple Nash equilibria, one may want to allow for randomizations between Nash equilibria by some form of communication prior to the play of the game.

Example 1.18   Battle of the Sexes   Suppose that in the Battle of the Sexes game, the players flip a coin and go to the Ballet if the outcome of the coin flip is heads, and to the Football game if the outcome is tails, i.e., they randomize between two pure strategy Nash equilibria, resulting in a payoff of (3/2, 3/2).

The coin flip is one way of communication prior to the play. A more general form of communication is to find a trusted mediator who can perform general randomizations, as illustrated in the next example.

Example 1.19   Traffic Intersection Game   Consider a game where two cars arrive at an intersection simultaneously. Row player (player 1) has the option to play U or D, and the column player (player 2) has the option to play L or R with payoffs given by

	L	R
U	5, 1	0, 0
D	4, 4	1, 5

Traffic Intersection Game.

There are two pure strategy Nash equilibria: the strategy profiles (U, L) and (D, R). To find the mixed strategy Nash equilibria, assume player 1 plays U with probability p and player 2 plays R with probability q. Using Proposition 1.15, we obtain

$$5q = 4q + (1 - q),$$
$$5p = 4p + (1 - p),$$

which imply that $p = q = 1/2$. This shows that there is a unique mixed strategy equilibrium with expected payoffs $(5/2, 5/2)$.

Case 1: Assume that there is a publicly observable random variable, e.g., a fair coin, such that if the outcome is heads, player 1 plays U and player 2 plays L, and if the outcome is tails, player 1 plays D and player 2 plays R. The expected payoffs for this play of the game is given by $(3, 3)$. We show that no player has an incentive to deviate from the "recommendation" of the coin. If player 1 sees heads, he believes that player 2 will play L, and, therefore, playing U is his best response (a similar argument holds when he sees tails). Similarly, if player 2 sees a heads, he believes that player 1 will play U, and, therefore, playing L is his best response (a similar argument holds when he sees tails).

Case 2: Next, consider a more elaborate signaling scheme. Suppose the players find a mediator who chooses $x \in \{1, 2, 3\}$ with equal probability $1/3$. She then sends the following messages:

- If $x = 1$, player 1 plays U, player 2 plays L.

- If $x = 2$, player 1 plays D, player 2 plays L.

- If $x = 3$, player 1 plays D, player 2 plays R.

  We show that no player has an incentive to deviate from the "recommendation" of the mediator:

- If player 1 gets the recommendation U, he believes player 2 will play L, so his best response is to play U.

- If player 1 gets the recommendation D, he believes player 2 will play L, R with equal probability, so his best response is to play D.

- If player 2 gets the recommendation L, he believes player 1 will play U, D with equal probability, so his best response is to play L.

- If player 2 gets the recommendation R, he believes player 1 will play D, so his best response is to play R.

  Thus, the players will follow the mediator's recommendations. With the mediator, the expected payoffs are $(10/3, 10/3)$, strictly higher than what the players could get by randomizing between Nash equilibria.

The preceding examples lead us to the notions of correlated strategies and "correlated equilibrium". Let $\Delta(S)$ denote the set of probability measures over a metric space S. Let R be a random variable taking values in $S = \Pi_{i=1}^{n} S_i$ distributed according to π. An instantiation of R is a pure strategy profile and the ith component of the instantiation will be called the *recommendation to player* i. Given such a recommendation, player i can use conditional probability to form posterior beliefs about the recommendations given to the other players. A distribution π is defined to be a *correlated equilibrium* if no player can ever expect to unilaterally gain by deviating from his recommendation, assuming the other players play according to their recommendations.

Definition 1.20   A *correlated equilibrium* of a finite game is a joint probability distribution $\pi \in \Delta(S)$ such that if R is a random variable distributed according to π then

$$\sum_{s_{-i} \in S_{-i}} \text{Prob}(R = s | R_i = s_i) \left[u_i(t_i, s_{-i}) - u_i(s) \right] \leq 0 \tag{1.1}$$

for all players i, all $s_i \in S_i$ such that $\text{Prob}(R_i = s_i) > 0$, and all $t_i \in S_i$.

We have the following useful characterization for correlated equilibria in finite games.

Proposition 1.21   *A joint distribution $\pi \in \Delta(S)$ is a correlated equilibrium of a finite game if and only if*

$$\sum_{s_{-i} \in S_{-i}} \pi(s) \left[u_i(t_i, s_{-i}) - u_i(s) \right] \leq 0 \tag{1.2}$$

for all players i and all $s_i, t_i \in S_i$ such that $s_i \neq t_i$.

Proof. Using the definition of conditional probability, we can rewrite the definition of a correlated equilibrium as

$$\sum_{s_{-i} \in S_{-i}} \frac{\pi(s)}{\sum_{t_{-i} \in S_{-i}} \pi(s_i, t_{-i})} \left[u_i(t_i, s_{-i}) - u_i(s) \right] \leq 0$$

for all i, all $s_i \in S_i$ such that $\sum_{t_{-i} \in S_{-i}} \pi(s_i, t_{-i}) > 0$, and all $t_i \in S_i$. The denominator does not depend on the variable of summation so it can be factored out of the sum and canceled, yielding the simpler condition that (1.2) holds for all i, all $s_i \in S_i$ such that $\sum_{t_{-i} \in S_{-i}} \pi(s_i, t_{-i}) > 0$, and all $t_i \in S_i$. But if $\sum_{t_{-i} \in S_{-i}} \pi(s_i, t_{-i}) = 0$ then the left-hand side of (1.2) is zero regardless of i and t_i, so the equation always holds trivially in this case. The equation (1.2) also holds trivially when $s_i = t_i$, so we only need to check it in the case $s_i \neq t_i$.   □

We can alternatively think of correlated equilibria as joint distributions corresponding to recommendations which will be given to the players as part of an extended game. The players are then free to play any function of their recommendation (this is called a *departure function*) as their strategy in the game. If it is a Nash equilibrium of this extended game for each player to play his recommended strategy (i.e., to use the identity departure function), then the distribution is a correlated equilibrium. This interpretation is justified by the following alternative characterization of correlated equilibria, which is useful in defining correlated equilibria for games with infinite strategy spaces (see [98]).

Proposition 1.22   *A joint distribution $\pi \in \Delta(S)$ is a correlated equilibrium of a finite game if and only if*

$$\sum_{s \in S} \pi(s) \left[u_i(\zeta_i(s_i), s_{-i}) - u_i(s) \right] \leq 0 \tag{1.3}$$

for all players i and all functions $\zeta_i : S_i \to S_i$.

Proof. By substituting $t_i = \zeta_i(s_i)$ into (1.2) and summing over all $s_i \in S_i$, we obtain (1.3) for any i and any $\zeta_i : S_i \to S_i$. For the converse, define ζ_i for any $s_i, t_i \in S_i$ by

$$\zeta_i(r_i) = \begin{cases} t_i & r_i = s_i \\ r_i & \text{else.} \end{cases}$$

Then all the terms in (1.3) except the s_i terms cancel, yielding (1.2). $\qquad \square$

1.3   EXISTENCE OF A NASH EQUILIBRIUM

In this section, we study the existence of a Nash equilibrium in both games with finite and infinite pure strategy sets. We start with an example *pricing-congestion game*, where players have infinitely many pure strategies. We consider two instances of this game, one with a unique pure Nash equilibrium, and the other with no pure Nash equilibria. This model will be studied in more detail in Chapter 3.

Example 1.23 Pricing-Congestion Game   We consider a price competition model that was studied in [5] (see also [4], [6], [83], and [27])[5].

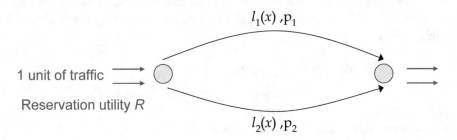

Figure 1.4: A price competition model over congested networks.

Consider a parallel link network with I links. Assume that d units of flow are to be routed through this network. We assume that this flow is the aggregate flow of many *infinitesimal* users.

[5]This model will be studied in detail in Chapter 3.

Let $l_i(x_i)$ denote the latency function of link i, which represents the delay or congestion cost as a function of the total flow x_i on link i. Assume that the links are owned by independent service providers, which set a price p_i per unit of flow on link i (see Figure 1.4). The effective cost of using link i is $p_i + l_i(x_i)$. Users have a reservation utility equal to R, i.e., if $p_i + l_i(x_i) > R$, then no traffic will be routed on link i.

We consider a special case of this model with two links and latency functions $l_1(x_1) = 0$ and $l_2(x_2) = \frac{3x_2}{2}$. For simplicity, we assume that $R = 1$ and $d = 1$. Given the prices (p_1, p_2), we assume that the flow is allocated according to a *Wardrop equilibrium* [37], i.e., the flows are routed along minimum effective cost paths and the effective cost cannot exceed the reservation utility (see also Def. 3.2 in Chapter 3). Formally, a flow vector $x = [x_i]_{i=1,...,I}$ is a Wardrop equilibrium if $\sum_{i=1}^{I} x_i \leq 1$ and

$$p_i + l_i(x_i) = \min_j\{p_j + l_j(x_j)\}, \qquad \text{for all } i \text{ with } x_i > 0,$$

$$p_i + l_i(x_i) \leq 1, \qquad \text{for all } i \text{ with } x_i > 0,$$

with $\sum_{i=1}^{I} x_i = 1$ if $\min_j\{p_j + l_j(x_j)\} < 1$.

We use the preceding characterization to determine the flow allocation on each link given prices $0 \leq p_1, p_2 \leq 1$:

$$x_2(p_1, p_2) = \begin{cases} \frac{2}{3}(p_1 - p_2), & p_1 \geq p_2, \\ 0, & \text{otherwise}, \end{cases}$$

and $x_1(p_1, p_2) = 1 - x_2(p_1, p_2)$. The payoffs for the providers are then given by:
$$u_1(p_1, p_2) = p_1 \times x_1(p_1, p_2),$$
$$u_2(p_1, p_2) = p_2 \times x_2(p_1, p_2).$$

We find the pure strategy Nash equilibria of this game by characterizing the best response correspondences, $B_i(p_{-i})$ for each player[6]. In particular, for a given p_2, $B_1(p_2)$ is the optimal solution set of the following optimization problem

$$\text{maximize}_{0 \leq p_1 \leq 1, \, 0 \leq x_1 \leq 1} \quad p_1 x_1$$
$$\text{subject to} \quad p_1 = p_2 + \frac{3}{2}(1 - x_1).$$

Solving the preceding optimization problem, we find that

$$B_1(p_2) = \min\left\{1, \frac{3}{4} + \frac{p_2}{2}\right\}.$$

Similarly, $B_2(p_1) = \frac{p_1}{2}$.

---

[6]The following analysis assumes that at the Nash equilibria (p_1, p_2) of the game, the corresponding Wardrop equilibria x satisfies $x_1 > 0, x_2 > 0$, and $x_1 + x_2 = 1$. For the proofs of these statements, see Acemoglu and Ozdaglar [5].

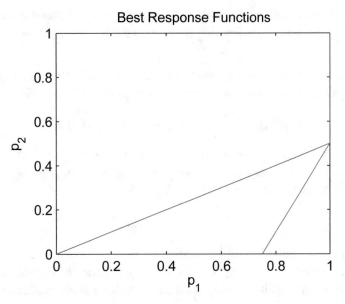

Figure 1.5: Best response correspondences.

Figure 1.5 illustrates the best response correspondences as a function of p_1 and p_2. These correspondences intersect at the unique point $(p_1, p_2) = (1, \frac{1}{2})$, which is the unique pure strategy Nash equilibrium.

We next consider a similar example with latency functions given by

$$l_1(x) = 0, \qquad l_2(x) = \begin{cases} 0 & \text{if } 0 \leq x \leq 1/2 \\ \frac{x - 1/2}{\epsilon} & x \geq 1/2, \end{cases}$$

for some sufficiently small $\epsilon > 0$. For this example, we show that no pure Nash equilibrium strategy exists for small ϵ. The following list considers all candidate Nash equilibria (p_1, p_2) and profitable unilateral deviations for ϵ sufficiently small, thus establishing the nonexistence of a pure strategy Nash equilibrium:

1. $p_1 = p_2 = 0$: A small increase in the price of provider 1 will generate positive profits, thus provider 1 has an incentive to deviate.

2. $p_1 = p_2 > 0$: Let x be the corresponding flow allocation. If $x_1 = 1$, then provider 2 has an incentive to decrease its price. If $x_1 < 1$, then provider 1 has an incentive to decrease its price.

3. $0 \leq p_1 < p_2$: Player 1 has an incentive to increase its price since its flow allocation remains the same.

4. $0 \leq p_2 < p_1$: For ϵ sufficiently small, the profit function of player 2, given p_1, is strictly increasing as a function of p_2, showing that provider 2 has an incentive to increase its price.

We are interested in establishing conditions which guarantee that a strategic form game has a (mixed) Nash equilibrium. Recall that a mixed strategy profile σ^* is a Nash equilibrium if

$$u_i(\sigma_i^*, \sigma_{-i}^*) \geq u_i(\sigma_i, \sigma_{-i}^*), \qquad \text{for all } \sigma_i \in \Sigma_i.$$

In other words, σ^* is a Nash equilibrium if and only if $\sigma_i^* \in B_i^*(\sigma_{-i}^*)$ for all i, where $B_i^*(\sigma_{-i}^*)$ is the best response of player i, given that the other players use the strategy profile σ_{-i}^*. We define the correspondence $B : \Sigma \rightrightarrows \Sigma$ such that for all $\sigma \in \Sigma$, we have

$$B(\sigma) = [B_i(\sigma_{-i})]_{i \in \mathcal{I}}. \tag{1.4}$$

The existence of a Nash equilibrium is then equivalent to the existence of a mixed strategy σ such that $\sigma \in B(\sigma)$, i.e., existence of a fixed point of the correspondence B. We will use Kakutani's fixed point theorem to establish conditions under which there exists a Nash equilibrium (see [53]).

Theorem 1.24 Kakutani's Fixed Point Theorem   *Let $f : A \rightrightarrows A$ be a correspondence, with $x \in A \to f(x) \subset A$, satisfying the following conditions:*

1. *A is a compact, convex, and non-empty subset of a finite dimensional Euclidean space.*

2. *$f(x)$ is non-empty for all $x \in A$.*

3. *$f(x)$ is a convex-valued correspondence: for all $x \in A$, $f(x)$ is a convex set.*

4. *$f(x)$ has a closed graph: If $\{x^n, y^n\} \to \{x, y\}$ with $y^n \in f(x^n)$, then $y \in f(x)$.*

Then, there exists some $x \in A$, such that $x \in f(x)$.

1.3.1   GAMES WITH FINITE PURE STRATEGY SETS

In this section, we show that a finite game always has a mixed strategy Nash equilibrium. In proving this, we use the following theorem that establishes the existence of an optimal solution for an optimization problem (see [21] for the proof).

Theorem 1.25 Weierstrass   *Let A be a nonempty compact subset of a finite dimensional Euclidean space and let $f : A \to \mathbb{R}$ be a continuous function. Then there exists an optimal solution to the optimization problem*

$$\begin{aligned} minimize \quad & f(x) \\ subject\ to \quad & x \in A. \end{aligned}$$

We proceed to the main equilibrium existence result, established by Nash [78].

Theorem 1.26   *Any finite strategic game has a mixed strategy Nash equilibrium.*

Proof. We will apply Kakutani's theorem to the best response correspondence $B : \Sigma \rightrightarrows \Sigma$ defined in Eq. (1.4). We show that $B(\sigma)$ satisfies the conditions of Kakutani's theorem.

1. The set Σ is compact, convex, and non-empty.

By definition, $\Sigma = \prod_{i \in \mathcal{I}} \Sigma_i$, where each Σ_i is a probability simplex of dimension $|S_i| - 1$, showing the desired property.

2. The set $B(\sigma)$ is non-empty for all $\sigma \in \Sigma$.

By definition,
$$B_i(\sigma_{-i}) \in \arg \max_{x \in \Sigma_i} u_i(x, \sigma_{-i}).$$

Here Σ_i is non-empty and compact, and u_i is linear in x. Therefore, Weirstrass' theorem applies, showing that $B(\sigma)$ is non-empty.

3. The correspondence $B(\sigma)$ is a convex-valued correspondence.

We show that for all $\sigma \in \Sigma$, the set $B(\sigma)$ is a convex set, or equivalently, $B_i(\sigma_{-i})$ is a convex set for all i. Let $\sigma'_i, \sigma''_i \in B_i(\sigma_{-i})$. Then, for all $\lambda \in [0, 1] \in B_i(\sigma_{-i})$, we have

$$u_i(\sigma'_i, \sigma_{-i}) \geq u_i(\tau_i, \sigma_{-i}) \qquad \text{for all } \tau_i \in \Sigma_i,$$

$$u_i(\sigma''_i, \sigma_{-i}) \geq u_i(\tau_i, \sigma_{-i}) \qquad \text{for all } \tau_i \in \Sigma_i.$$

The preceding relations imply that for all $\lambda \in [0, 1]$, we have

$$\lambda u_i(\sigma'_i, \sigma_{-i}) + (1 - \lambda) u_i(\sigma''_i, \sigma_{-i}) \geq u_i(\tau_i, \sigma_{-i}) \qquad \text{for all } \tau_i \in \Sigma_i.$$

By the linearity of u_i,

$$u_i(\lambda \sigma'_i + (1 - \lambda) \sigma''_i, \sigma_{-i}) \geq u_i(\tau_i, \sigma_{-i}) \qquad \text{for all } \tau_i \in \Sigma_i.$$

Therefore, $\lambda \sigma'_i + (1 - \lambda) \sigma''_i \in B_i(\sigma_{-i})$, showing that $B(\sigma)$ is convex-valued.

4. The correspondence $B(\sigma)$ has a closed graph.

Assume, to arrive at a contradiction, that $B(\sigma)$ does not have a closed graph. Then, there exists a sequence $(\sigma^n, \hat{\sigma}^n) \to (\sigma, \hat{\sigma})$ with $\hat{\sigma}^n \in B(\sigma^n)$, but $\hat{\sigma} \notin B(\sigma)$, i.e., there exists some i such that $\hat{\sigma}_i \notin B_i(\sigma_{-i})$. This implies that there exists some $\sigma'_i \in \Sigma_i$ and some $\epsilon > 0$ such that

$$u_i(\sigma'_i, \sigma_{-i}) > u_i(\hat{\sigma}_i, \sigma_{-i}) + 3\epsilon.$$

By the continuity of u_i and the fact that $\sigma^n_{-i} \to \sigma_{-i}$, we have for sufficiently large n,

$$u_i(\sigma'_i, \sigma^n_{-i}) \geq u_i(\sigma'_i, \sigma_{-i}) - \epsilon.$$

Combining the preceding two relations, we obtain

$$u_i(\sigma'_i, \sigma^n_{-i}) > u_i(\hat{\sigma}_i, \sigma_{-i}) + 2\epsilon \geq u_i(\hat{\sigma}^n_i, \sigma^n_{-i}) + \epsilon,$$

where the second relation follows from the continuity of u_i. This contradicts the assumption that $\hat{\sigma}^n_i \in B_i(\sigma^n_{-i})$, and completes the proof.   □

Since every mixed Nash equilibrium is a correlated equilibrium, the existence of a correlated equilibrium in a finite strategic form game follows from this result.

1.3.2   GAMES WITH INFINITE PURE STRATEGY SETS

In the previous section, we showed that every finite strategic form game has a mixed strategy Nash equilibrium. The proof relies on Kakutani's fixed point theorem and the fact that using mixed strategies essentially "convexifies" the strategy spaces in finite games. In this section, we investigate the question of the existence of a pure Nash equilibrium in games where the players have infinitely many pure strategies. In particular, we show that under convexity assumptions on the strategy sets and the payoff functions, there exists a pure strategy Nash Equilibrium.

We first introduce some basic convexity concepts. For a given set $C \subset \mathbb{R}^n$ and function $f : C \to \mathbb{R}$, denote the *t-upper level set* of f by

$$L_f(t) = \{x \in C \mid f(x) \geq t\}.$$

Definition 1.27   Let C be a nonempty convex subset of $\mathbb{R}^n$.

(i) A function $g : C \to \mathbb{R}$ is called *concave* if for all $x, y \in C$, and for all $\lambda \in [0, 1]$, we have

$$f(\lambda x + (1 - \lambda)y) \geq \lambda f(x) + (1 - \lambda)f(y).$$

(ii) A function $g : C \to \mathbb{R}$ is called *quasi-concave* if for all t, $L_g(t)$ is a convex set, i.e., for all $x, y \in C$ with $g(x) \geq t$ and $g(y) \geq t$, and for all $\lambda \in [0, 1]$, we have

$$g(\lambda x + (1 - \lambda)y) \geq t.$$

We next show that under convexity assumptions, games with infinite strategy sets have pure Nash equilibria.

Theorem 1.28   Debreu, Glicksberg and Fan   *Consider a strategic form game $\langle \mathcal{I}, (S_i), (u_i) \rangle$, where $\mathcal{I}$ is a finite set. Assume that the following holds for each $i \in \mathcal{I}$:*

1. S_i is a non-empty, convex, and compact subset of a finite-dimensional Euclidean space.

2. $u_i(s)$ is continuous in s.

3. $u_i(s_i, s_{-i})$ is quasi-concave in s_i.

Then, the game $\langle \mathcal{I}, (s_i), (u_i) \rangle$ has a pure strategy Nash equilibrium.

Proof. (outline) Consider the best response correspondence $B : S \rightrightarrows S$, where $S = \prod_{i=1}^{I} S_i$, defined by

$$B(s) = [B_i(s_{-i})]_{i \in \mathcal{I}}.$$

Following a similar argument as in the proof of Theorem 1.26, it can be seen that the correspondence B satisfies the conditions of Kakutani's theorem: in step 1, use the convexity assumption on the strategy sets, and in step 3, use the quasi-concavity assumption on the utility functions instead of linearity.   □

The existence of a mixed Nash equilibrium in finite games is a special case of the preceding theorem: for finite games, the utility function of a player is linear in its mixed strategy, hence satisfies the assumptions of the preceding theorem. What happens when we relax quasi-concavity? The following example shows that without this assumption, a pure strategy Nash equilibrium may fail to exist.

Example 1.29   Unit Circle Example

Two players pick points s_1 and s_2 on the unit circle. The payoffs for the two players are

$$u_1(s_1, s_2) = d(s_1, s_2)$$

$$u_2(s_1, s_2) = -d(s_1, s_2)$$

where $d(x, y)$ denotes the Euclidean distance between the vectors $x, y \in \mathbb{R}^2$. It can be seen that there is no pure strategy Nash equilibrium in this game, i.e., if both players pick the same location, player 1 has an incentive to deviate. If they pick different locations, player 2 has an incentive to deviate.

Despite the non-existence of a pure strategy Nash equilibrium in the preceding example, one can show that each player choosing a uniformly random point on the circle is a mixed strategy Nash equilibrium. In the next section, we will address the question of existence of a mixed Nash equilibrium in games where each player has infinitely many pure strategies. We will first study "continuous games". We will then focus on "discontinuous games", which arise in competition models among service providers in communication networks, and delineate the conditions under which a mixed Nash equilibrium exists.

1.3.3   CONTINUOUS GAMES

In this section, we study games in which players have infinitely many pure strategies. In particular, we want to include the possibility that the pure strategy set of a player may be a bounded interval on the real line, such as $[0,1]$.[7]

Definition 1.30   A *continuous game* is a game $\langle \mathcal{I}, (S_i), (u_i) \rangle$ where $\mathcal{I}$ is a finite set, the S_i are nonempty compact metric spaces, and the $u_i : S \to \mathbb{R}$ are continuous functions.

A *compact metric space* is a general mathematical structure for representing infinite sets that can be well approximated by large finite sets. One important fact is that, in a compact metric space, any infinite sequence has a convergent subsequence. Any closed bounded subset of a finite-dimensional Euclidean space is an example of a compact metric space. More specifically, any closed bounded interval of the real line is an example of a compact metric space, where the distance between two points x and y is given by $|x - y|$. In our treatment, we will not need to refer to any examples more complicated than these (see Appendix 1.A for basic definitions of a metric space and convergence notions for probability measures).

We next state the analogue of Nash's Theorem for continuous games.

Theorem 1.31   Glicksberg   *Every continuous game has a mixed strategy Nash equilibrium.*

With continuous strategy spaces, the space of mixed strategies Σ is infinite-dimensional; therefore, we need a more powerful fixed point theorem than the version of Kakutani we have used before. Here we adopt an alternative approach to prove Glicksberg's Theorem, which can be summarized as follows:

- We approximate the original game with a sequence of finite games, which correspond to successively finer discretizations of the original game.

- We use Nash's Theorem to produce an equilibrium for each approximation.

- We use the weak topology and the continuity assumptions to show that these converge to an equilibrium of the original game.

1.3.3.1   Closeness of Two Games and ϵ-Equilibrium

Let $u = (u_1, \ldots, u_I)$ and $\tilde{u} = (\tilde{u}_1, \ldots, \tilde{u}_I)$ be two profiles of utility functions defined on S such that for each $i \in \mathcal{I}$, the functions $u_i : S \to \mathbb{R}$ and $\tilde{u}_i : S \to \mathbb{R}$ are bounded measurable functions. We may define the distance between the utility function profiles u and $\tilde{u}$ as

$$\max_{i \in \mathcal{I}} \sup_{s \in S} |u_i(s) - \tilde{u}_i(s)|.$$

---

[7]Our development follows that of Myerson [77].

Consider two strategic form games defined by the two profiles of utility functions:

$$G = \langle \mathcal{I}, (S_i), (u_i) \rangle, \qquad \tilde{G} = \langle \mathcal{I}, (S_i), (\tilde{u}_i) \rangle.$$

Even if u and $\tilde{u}$ are very close, the equilibria of G and $\tilde{G}$ may be far apart. For example, assume there is only one player, $S_1 = [0, 1]$, $u_1(s_1) = \epsilon s_1$, and $\tilde{u}_1(s_1) = -\epsilon s_1$, where $\epsilon > 0$ is a sufficiently small scalar. The unique equilibrium of G is $s_1^* = 1$, and the unique equilibrium of $\tilde{G}$ is $s_1^* = 0$, even if the distance between u and $\tilde{u}$ is only 2ϵ.

However, if u and $\tilde{u}$ are very close, there is a sense in which the equilibria of G are "almost" equilibria of $\tilde{G}$.

Definition 1.32 (ϵ-equilibrium) Given $\epsilon \geq 0$, a mixed strategy $\sigma \in \Sigma$ is called an ϵ-equilibrium if for all $i \in \mathcal{I}$ and $s_i \in S_i$,

$$u_i(s_i, \sigma_{-i}) \leq u_i(\sigma_i, \sigma_{-i}) + \epsilon.$$

Clearly, an ϵ-equilibrium with $\epsilon = 0$ is a Nash equilibrium.

The following result shows that such ϵ-equilibria have a continuity property across games.

Proposition 1.33 *Let G be a continuous game. Assume that $\sigma^k \to \sigma$, $\epsilon^k \to \epsilon$, and for each k, σ^k is an ϵ^k-equilibrium of G. Then σ is an ϵ-equilibrium of G.*

Proof. For all $i \in \mathcal{I}$, and all $s_i \in S_i$, we have

$$u_i(s_i, \sigma_{-i}^k) \leq u_i(\sigma^k) + \epsilon.$$

Taking the limit as $k \to \infty$ in the preceding relation, and using the continuity of the utility functions together with the convergence of probability distributions under weak topology [see Eq. (1.14)], we obtain,

$$u_i(s_i, \sigma_{-i}) \leq u_i(\sigma) + \epsilon,$$

establishing the result. $\qquad \qquad \square$

We next define formally the closeness of two strategic form games.

Definition 1.34 Let G and G' be two strategic form games with

$$G = \langle \mathcal{I}, (S_i), (u_i) \rangle, \qquad \tilde{G} = \langle \mathcal{I}, (S_i), (\tilde{u}_i) \rangle.$$

Assume that the utility functions u_i and $\tilde{u}_i$ are measurable and bounded. Then G' is an $\alpha-$approximation to G if for all $i \in \mathcal{I}$ and $s \in S$, we have

$$|u_i(s) - u_i'(s)| \leq \alpha.$$

We next relate the ϵ-equilibria of close games.

Proposition 1.35   *If G' is an α-approximation to G and σ is an ϵ-equilibrium of G', then σ is an $(\epsilon + 2\alpha)$-equilibrium of G.*

Proof. For all $i \in \mathcal{I}$ and all $s_i \in S_i$, we have

$$
\begin{aligned}
u_i(s_i, \sigma_{-i}) - u_i(\sigma) &= u_i(s_i, \sigma_{-i}) - u_i'(s_i, \sigma_{-i}) + u_i'(s_i, \sigma_{-i}) - u_i'(\sigma) + u_i'(\sigma) - u_i(\sigma) \\
&\leq \epsilon + 2\alpha.
\end{aligned}
$$

$\square$

The next proposition shows that we can approximate a continuous game with an essentially finite game to an arbitrary degree of accuracy.

Proposition 1.36   *For any continuous game G and any $\alpha > 0$, there exists an "essentially finite" game which is an α-approximation to G.*

Proof. Since S is a compact metric space, the utility functions u_i are uniformly continuous, i.e., for all $\alpha > 0$, there exists some $\epsilon > 0$ such that

$$
u_i(s) - u_i(t) \leq \alpha,
$$

for all $d(s, t) \leq \epsilon$. Since S_i is a compact metric space, it can be covered with finitely many open balls U_i^j, each with radius less than ϵ. Assume without loss of generality that these balls are disjoint and nonempty. Choose an $s_i^j \in U_i^j$ for each i, j. Define the "essentially finite" game G' with the utility functions u_i' defined as

$$
u_i'(s) = u_i(s_1^j, \ldots, s_I^j), \qquad \text{for all } s \in U^j = \prod_{k=1}^{I} U_k^j.
$$

Then for all $s \in S$ and all $i \in \mathcal{I}$, we have

$$
|u_i'(s) - u_i(s)| \leq \alpha,
$$

since $d(s, s^j) \leq \epsilon$ for all j, implying the desired result.

$\square$

1.3.3.2 Proof of Glicksberg's Theorem

We now return to the proof of Glicksberg's Theorem. Let $\{\alpha^k\}$ be a scalar sequence with $\alpha^k \downarrow 0$.

- For each α^k, there exists an "essentially finite" α^k-approximation G^k of G by Proposition 1.36.

- Since G^k is "essentially finite" for each k, it follows using Nash's Theorem that it has a 0-equilibrium, which we denote by σ^k.

- Then, by Proposition 1.35, σ^k is a $2\alpha^k$-equilibrium of G.

- Since Σ is compact, $\{\sigma^k\}$ has a convergent subsequence. Without loss of generality, we assume that $\sigma^k \to \sigma$.

- Since $2\alpha^k \to 0$, $\sigma^k \to \sigma$, by Proposition 1.33, it follows that σ is a 0-equilibrium of G.

1.3.4 DISCONTINUOUS GAMES

There are many games in which the utility functions are not continuous (e.g., price competition models, congestion-competition models in networks). The next theorem shows that for discontinuous games, under some mild semicontinuity conditions on the utility functions, it is possible to establish the existence of a mixed Nash equilibrium (see Dasgupta and Maskin [40]-[41]). The key assumption is to allow discontinuities in the utility function to occur only on a subset of measure zero, in which a player's strategy is "related" to another player's strategy. To formalize this notion, we introduce the following set: for any two players i and j, let D be a finite index set and for $d \in D$, let $f_{ij}^d : S_i \to S_j$ be a bijective and continuous function. Then, for each i, we define

$$S^*(i) = \{s \in S \mid \text{there exists } j \neq i \text{ such that } s_j = f_{ij}^d(s_i).\} \qquad (1.5)$$

Before stating the theorem, we first introduce some weak continuity conditions.

Definition 1.37   Let X be a subset of $\mathbb{R}^n$, X_i be a subset of $\mathbb{R}$, and X_{-i} be a subset of $\mathbb{R}^{n-1}$.

(i) A function $f : X \to \mathbb{R}$ is called *upper semicontinuous* (respectively, *lower semicontinuous*) at a vector $x \in X$ if $f(x) \geq \limsup_{k \to \infty} f(x_k)$ (respectively, $f(x) \leq \liminf_{k \to \infty} f(x_k)$) for every sequence $\{x_k\} \subset X$ that converges to x. If f is upper semicontinuous (lower semicontinuous) at every $x \in X$, we say that f is upper semicontinuous (lower semicontinuous).

(ii) A function $f : X_i \times X_{-i} \to \mathbb{R}$ is called *weakly lower semicontinuous* in x_i over a subset $X_{-i}^* \subset X_{-i}$, if for all x_i there exists $\lambda \in [0, 1]$ such that, for all $x_{-i} \in X_{-i}^*$,

$$\lambda \liminf_{x_i' \uparrow x_i} f(x_i', x_{-i}) + (1 - \lambda) \liminf_{x_i' \downarrow x_i} f(x_i', x_{-i}) \geq f(x_i, x_{-i}).$$

Theorem 1.38   *(Dasgupta and Maskin) Let S_i be a closed interval of $\mathbb{R}$. Assume that u_i is continuous except on a subset $S^{**}(i)$ of the set $S^*(i)$ defined in Eq. (1.5). Assume also that $\sum_{i=1}^{n} u_i(s)$ is upper semicontinuous and that $u_i(s_i, s_{-i})$ is bounded and weakly lower semicontinuous in s_i over the set $\{s_{-i} \in S_{-i} \mid (s_i, s_{-i}) \in S^{**}(i)\}$. Then the game has a mixed strategy Nash equilibrium.*

The weakly lower semicontinuity condition on the utility functions implies that the function u_i does not jump up when approaching s_i, either from below or above. Loosely, this ensures that player i can do almost as well with strategies near s_i as with s_i, even if his opponents put weight on the discontinuity points of u_i.

In the following examples, we consider games with discontinuous utility functions and study their pure and mixed strategy Nash equilibria. The first example is the Bertrand competition, which is a standard model of competition among firms selling a homogeneous good. It is particularly relevant in network games where service providers set prices over resources and compete over user demand.

Example 1.39   Bertrand Competition with Capacity Constraints   Consider two firms that charge prices $p_1, p_2 \in [0, 1]$ per unit of the same good. Assume that there is unit demand and all customers choose the firm with the lower price. If both firms charge the same price, each firm gets half the demand. All demand has to be supplied. The payoff functions of each firm is the profit they make (we assume for simplicity that cost of supplying the good is equal to 0 for both firms).

(a) We first study the pure strategy Nash equilibria of this game. For this, we consider all possible candidate strategy profiles and check if there is any profitable unilateral deviation:

   - $p_1 = p_2 > 0$: each of the firms has an incentive to reduce their price to capture the whole demand and increase profits.

   - $p_1 < p_2$: Firm 1 has an incentive to slightly increase his price.

   - $p_1 = p_2 = 0$: Neither firm can increase profits by changing its price unilaterally. Hence, $(p_1, p_2) = (0, 0)$ is the unique pure strategy Nash equilibrium.

(b) Assume now that each firm has a capacity constraint of 2/3 units of demand (since all demand has to be supplied, this implies that when $p_1 < p_2$, firm 2 gets 1/3 units of demand). It can be seen in this case that the strategy profile $(p_1, p_2) = (0, 0)$ is no longer a pure strategy Nash equilibrium: either firm can increase his price and still have 1/3 units of demand due to the capacity constraint on the other firm, thus making positive profits.

   It can be established using Theorem 1.38 that there exists a mixed strategy Nash equilibrium. Let us next proceed to construct a mixed strategy Nash equilibrium. We focus on symmetric Nash equilibria, i.e., both firms use the same mixed strategy. We use the cumulative distribution function $F(\cdot)$ to represent the mixed strategy used by either firm. It can be seen that the

expected payoff of player 1, when he chooses p_1 and firm 2 uses the mixed strategy $F(\cdot)$, is given by

$$u_1(p_1, F(\cdot)) = F(p_1)\frac{p_1}{3} + (1 - F(p_1))\frac{2}{3}p_1.$$

Using the fact that each action in the support of a mixed strategy must yield the same payoff to a player at the equilibrium (cf. Proposition 1.15), we obtain for all p in the support of $F(\cdot)$,

$$-F(p)\frac{p}{3} + \frac{2}{3}p = k,$$

for some $k \geq 0$. From this we obtain:

$$F(p) = 2 - \frac{3k}{p}.$$

Note next that the upper support of the mixed strategy must be at $p = 1$, which implies that $F(1) = 1$. Combining with the preceding, we obtain

$$F(p) = \begin{cases} 0, & \text{if } 0 \leq p \leq \frac{1}{2}, \\ 2 - \frac{1}{p}, & \text{if } \frac{1}{2} \leq p \leq 1, \\ 1, & \text{if } p \geq 1. \end{cases}$$

The analysis of the equilibria of the next game is similar and left as an exercise.

Example 1.40 Hoteling Competition   Each of n candidates chooses a position to take on the real line in the interval [0,1]. There is a continuum of citizens, whose favorite positions are uniformly distributed between [0,1]. A candidate attracts votes of citizens whose favorite positions are closer to his position than to the position of any other candidate; if k candidates choose the same position, then each receives the fraction $1/k$ of the votes that the position attracts. The payoff of each candidate is his vote share.

(a) Find all pure strategy Nash equilibria when $n = 2$.

(b) Show that there does not exist a pure strategy Nash equilibrium when $n = 3$. Find a mixed strategy Nash equilibrium.

1.4   UNIQUENESS OF A NASH EQUILIBRIUM

In the previous section, we showed that under some convexity assumptions, games with infinite pure strategy sets have a pure Nash equilibrium (see Theorem 1.28). The next example shows that even under strict convexity assumptions, there may be infinitely many pure strategy Nash equilibria.

Example 1.41   Consider a game with 2 players, where $S_i = [0, 1]$ for $i = 1, 2$, and the payoffs are given by

$$u_1(s_1, s_2) = s_1 s_2 - \frac{s_1^2}{2},$$

$$u_2(s_1, s_2) = s_1 s_2 - \frac{s_2^2}{2}.$$

Note that $u_i(s_1, s_2)$ is strictly concave in s_i. It can be seen in this example that the best response correspondences (which are unique-valued) are given by

$$B_1(s_2) = s_2, \qquad B_2(s_1) = s_1.$$

Plotting the best response curves shows that any pure strategy profile $(s_1, s_2) = (x, x)$ for $x \in [0, 1]$ is a pure strategy Nash equilibrium.

We will next establish conditions that guarantee that a strategic form game has a unique pure strategy Nash equilibrium. We will follow the development of the classical paper by Rosen [88].

In order to discuss the uniqueness of an equilibrium, we provide a more explicit description of the strategy sets of the players. In particular, we assume that for player $i \in \mathcal{I}$, the strategy set S_i is given by

$$S_i = \{x_i \in \mathbb{R}^{m_i} \mid h_i(x_i) \geq 0\}, \tag{1.6}$$

where $h_i : \mathbb{R}^{m_i} \rightarrow \mathbb{R}$ is a concave function. Since h_i is concave, it follows that the set S_i is a convex set. Therefore, the set of strategy profiles $S = \prod_{i=1}^{I} S_i \subset \prod_{i=1}^{I} \mathbb{R}^{m_i}$, being a Cartesian product of convex sets, is a convex set. The following analysis can be extended to the case where S_i is represented by finitely many concave inequality constraints, but we do not do so here for clarity of exposition.

Given these strategy sets, a vector $x^* \in \prod_{i=1}^{I} \mathbb{R}^{m_i}$ is a pure strategy Nash equilibrium if and only if for all $i \in \mathcal{I}$, x_i^* is an optimal solution of the optimization problem

$$\begin{aligned} \text{maximize}_{y_i \in \mathbb{R}^{m_i}} \quad & u_i(y_i, x_{-i}^*) \\ \text{subject to} \quad & h_i(y_i) \geq 0. \end{aligned} \tag{1.7}$$

(for a brief review of standard notation and results for nonlinear optimization problems, see Appendix 1.B). In the following, we use the notation $\nabla u(x)$ for the gradient vector, namely

$$\nabla u(x) = [\nabla_1 u_1(x), \ldots, \nabla_I u_I(x)]^T. \tag{1.8}$$

We next introduce the key condition for uniqueness of a pure strategy Nash equilibrium.

Definition 1.42 We say that the payoff functions $(u_1, \ldots, u_I)$ are *diagonally strictly concave for* $x \in S$, if for every $x^*, \bar{x} \in S$, we have

$$(\bar{x} - x^*)^T \nabla u(x^*) + (x^* - \bar{x})^T \nabla u(\bar{x}) > 0.$$

Theorem 1.43 *Consider a strategic form game* $\langle \mathcal{I}, (S_i), (u_i) \rangle$. *For all* $i \in \mathcal{I}$, *assume that the strategy sets* S_i *are given by Eq.* (1.6), *where* h_i *is a concave function, and there exists some* $\tilde{x}_i \in \mathbb{R}^{m_i}$ *such that* $h_i(\tilde{x}_i) > 0$. *Assume also that the payoff functions* $(u_1, \ldots, u_I)$ *are diagonally strictly concave for* $x \in S$. *Then the game has a unique pure strategy Nash equilibrium.*

Proof. Assume that there are two distinct pure strategy Nash equilibria. Since for each $i \in \mathcal{I}$, both x_i^* and $\bar{x}_i$ must be an optimal solution for an optimization problem of the form (1.7), Theorem 1.46 implies the existence of nonnegative vectors $\lambda^* = [\lambda_1^*, \ldots, \lambda_I^*]^T$ and $\bar{\lambda} = [\bar{\lambda}_1, \ldots, \bar{\lambda}_I]^T$ such that for all $i \in \mathcal{I}$, we have

$$\nabla_i u_i(x^*) + \lambda_i^* \nabla h_i(x_i^*) = 0, \tag{1.9}$$

$$\lambda_i^* h_i(x_i^*) = 0, \tag{1.10}$$

and

$$\nabla_i u_i(\bar{x}) + \bar{\lambda}_i \nabla h_i(\bar{x}_i) = 0, \tag{1.11}$$

$$\bar{\lambda}_i h_i(\bar{x}_i) = 0. \tag{1.12}$$

Multiplying Eqs. (1.9) and (1.11) by $(\bar{x}_i - x_i^*)^T$ and $(x_i^* - \bar{x}_i)^T$, respectively, and adding over all $i \in \mathcal{I}$, we obtain

$$
\begin{aligned}
0 = (\bar{x} - x^*)^T \nabla u(x^*) \;+\;& (x^* - \bar{x})^T \nabla u(\bar{x}) \\
+\;& \sum_{i \in \mathcal{I}} \lambda_i^* \nabla h_i(x_i^*)^T (\bar{x}_i - x_i^*) + \sum_{i \in \mathcal{I}} \bar{\lambda}_i \nabla h_i(\bar{x}_i)^T (x_i^* - \bar{x}_i) \\
>\;& \sum_{i \in \mathcal{I}} \lambda_i^* \nabla h_i(x_i^*)^T (\bar{x}_i - x_i^*) + \sum_{i \in \mathcal{I}} \bar{\lambda}_i \nabla h_i(\bar{x}_i)^T (x_i^* - \bar{x}_i), \tag{1.13}
\end{aligned}
$$

where to get the strict inequality, we used the assumption that the payoff functions are diagonally strictly concave for $x \in S$. Since the h_i are concave functions, we have

$$h_i(x_i^*) + \nabla h_i(x_i^*)^T (\bar{x}_i - x_i^*) \geq h_i(\bar{x}_i).$$

Using the preceding together with $\lambda_i^* > 0$, we obtain for all i,

$$
\begin{aligned}
\lambda_i^* \nabla h_i(x_i^*)^T (\bar{x}_i - x_i^*) &\geq \lambda_i^* (h_i(\bar{x}_i) - h_i(x_i^*)) \\
&= \lambda_i^* h_i(\bar{x}_i) \\
&\geq 0,
\end{aligned}
$$

where to get the equality we used Eq. (1.10), and to get the last inequality, we used the facts $\lambda_i^* \geq 0$ and $h_i(\bar{x}_i) \geq 0$. Similarly, we have

$$
\bar{\lambda}_i \nabla h_i(\bar{x}_i)^T (x_i^* - \bar{x}_i) \geq 0.
$$

Combining the preceding two relations with the relation in (1.13) yields a contradiction, thus concluding the proof. □

Let $U(x)$ denote the *Jacobian* of $\nabla u(x)$ [see Eq. (1.8)]. In particular, if the x_i are all 1-dimensional, then $U(x)$ is given by

$$
U(x) = \begin{pmatrix}
\frac{\partial^2 u_1(x)}{\partial x_1^2} & \frac{\partial^2 u_1(x)}{\partial x_1 \partial x_2} & \cdots \\
\frac{\partial^2 u_2(x)}{\partial x_2 \partial x_1} & \ddots & \\
\vdots & &
\end{pmatrix}.
$$

The next proposition provides a sufficient condition for the payoff functions to be diagonally strictly concave.

Proposition 1.44   *For all $i \in \mathcal{I}$, assume that the strategy sets S_i are given by Eq. (1.6), where h_i is a concave function. Assume that the symmetric matrix $(U(x) + U^T(x))$ is negative definite for all $x \in S$, i.e., for all $x \in S$, we have*

$$
y^T (U(x) + U^T(x))y < 0, \qquad \forall\, y \neq 0.
$$

Then, the payoff functions $(u_1, \ldots, u_I)$ are diagonally strictly concave for $x \in S$.

Proof. Let $x^*,\ \bar{x} \in S$. Consider the vector

$$
x(\lambda) = \lambda x^* + (1 - \lambda)\bar{x}, \qquad \text{for some } \lambda \in [0, 1].
$$

Since S is a convex set, $x(\lambda) \in S$.

Because $U(x)$ is the Jacobian of $\nabla u(x)$, we have

$$
\begin{aligned}
\frac{d}{d\lambda} \nabla u(x(\lambda)) &= U(x(\lambda)) \frac{dx(\lambda)}{d(\lambda)} \\
&= U(x(\lambda))(x^* - \bar{x}),
\end{aligned}
$$

or

$$\int_0^1 U(x(\lambda))(x^* - \bar{x})d\lambda = \nabla u(x^*) - \nabla u(\bar{x}).$$

Multiplying the preceding by $(\bar{x} - x^*)^T$ yields

$$
\begin{aligned}
(\bar{x} - x^*)^T \nabla u(x^*) \quad &+ \quad (x^* - \bar{x})^T \nabla u(\bar{x}) \\
&= -\frac{1}{2} \int_0^1 (x^* - \bar{x})^T [U(x(\lambda)) + U^T(x(\lambda))](x^* - \bar{x})d\lambda \\
&> 0,
\end{aligned}
$$

where to get the strict inequality we used the assumption that the symmetric matrix $(U(x) + U^T(x))$ is negative definite for all $x \in S$. $\square$

Recent work has shown that the uniqueness of a pure strategy Nash equilibrium can be established under conditions that are weaker than those given above. This analysis is beyond our scope; for more information on this, see the papers [96], [97], [95].

1.A APPENDIX: METRIC SPACES AND PROBABILITY MEASURES

This appendix summarizes the basic definitions related to metric spaces and probability measures for completeness. A *metric space* is a set M together with a function $d : M \times M \to \mathbb{R}$, that defines the "distance" $d(x, y)$ between any two points x, y in the set. The distance function satisfies the following properties for every $x, y, z \in M$:

- $d(x, y) = d(y, x) \geq 0$,

- $d(x, y) = 0$ if and only if $x = y$,

- $d(x, y) + d(y, z) \geq d(x, z)$.

In a metric space M, a point $y \in M$ is the limit of a sequence of points $\{x^k\}_{k=1}^{\infty} \subset M$ if and only if the distance $d(x^k, y) \to 0$ as $k \to \infty$. An *open ball* of radius ϵ around a point x, denoted by $B(x, \epsilon)$, is the set of all points in the metric space that have a distance less than ϵ from x, i.e.,

$$B(x, \epsilon) = \{y \mid d(x, y) < \epsilon\}.$$

A set S is an *open* subset of a metric space M if and only if for every $x \in S$, there exists some $\epsilon > 0$ such that $B(x, \epsilon) \subset S$. A metric space is compact if and only if every collection of open sets that "covers" M (i.e., their union includes all of M) has a finite sub-collection that also covers M.

When there are infinitely many actions in the set S_i, a mixed strategy for player i can no longer be described by just listing the probability of each individual action (i.e., by a finite dimensional probability vector). For example, suppose that S_i is the interval $[0, 1]$. If player i selected his action from a uniform probability distribution over the interval [0,1], then each individual action in [0,1] would have zero probability; but the same would be true if he selected his action from a uniform probability distribution over the interval [0.5,1]. To describe a probability distribution over S_i, we must list the probabilities of subsets of S_i. Unfortunately, for technical reasons, it may be mathematically impossible to consistently assign probabilities to all subsets of an infinite set, so some weak restriction is needed on the class of subsets whose probabilities can be meaningfully defined. These are called the *measurable sets*. Here, we let the measurable subsets of S and of each set S_i be the smallest class of subsets that includes all open subsets, all closed subsets, and all finite or countably infinite unions and intersections of sets in the class. These are the *Borel subsets* (and they include essentially all subsets that could be defined without the use of very sophisticated mathematics). Let $\mathcal{B}_i$ denote the set of such measurable or Borel subsets of S_i.

Let Σ_i denote the set of probability distributions over S_i, i.e., $\sigma_i \in \Sigma_i$ if and only if σ_i is a function that assigns a nonnegative number $\sigma_i(Q)$ to each Q that is a Borel subset of S_i, $\sigma(S_i) = 1$, and for any countable collection $(Q^k)_{k=1}^{\infty}$ of pairwise-disjoint Borel subsets of S_i,

$$\sigma_i \left(\bigcup_{k=1}^{\infty} Q^k \right) = \sum_{k=1}^{\infty} \sigma_i(Q^k).$$

We define convergence for mixed strategies by assigning the *weak topology* to Σ_i. Two implications of this topology are important for our purposes:

- The set of probability distributions Σ_i is a compact metric space, i.e., every sequence $\{\sigma_i^k\} \subset \Sigma_i$ has a convergent subsequence.

- A sequence $\{\sigma_i^k\} \subset \Sigma_i$ of mixed strategies converges to $\sigma_i \in \Sigma_i$ if and only if for all continuous functions $f_i : S_i \to \mathbb{R}$, we have

$$\lim_{k \to \infty} \int_{S_i} f(s_i) d\sigma_i^k(s_i) = \int_{S_i} f(s_i) d\sigma_i(s_i). \tag{1.14}$$

The preceding condition asserts that if $\tilde{s}_i^k$ is a random strategy drawn from S_i according to the σ_i^k distribution, and $\tilde{s}_i$ is a random strategy drawn from S_i according to the σ_i distribution, then the expected value of $f(\tilde{s}_i^k)$ must converge to the expected value of $f(\tilde{s}_i)$ as $k \to \infty$.

A function $g : S \to \mathbb{R}$ is called *(Borel) measurable* if for every scalar t, the set $\{x \in S \mid g(x) \geq t\}$ is a Borel subset of S. A function $g : S \to \mathbb{R}$ is bounded if there exists some scalar K such that $|g(x)| \leq K$ for all $x \in S$. To be able to define expected utilities, we must require that player's utility functions are measurable and bounded in this sense. Note that this assumption is weaker than the continuity of the utility functions. With this assumption, the utility functions u_i are extended from $S = \prod_{j=1}^{I} S_j$ to the space of probability distributions $\Sigma = \prod_{j=1}^{I} \Sigma_j$ as follows:

$$u_i(\sigma) = \int_S u_i(s) d\sigma(s),$$

where $\sigma \in \Sigma$.

1.B   APPENDIX: NONLINEAR OPTIMIZATION

This appendix provides some notation and standard results for nonlinear optimization problems.

Given a scalar-valued function $f : \mathbb{R}^n \to \mathbb{R}$, we use the notation $\nabla f(x)$ to denote the gradient vector of f at point x, i.e.,

$$\nabla f(x) = \left[\frac{\partial f(x)}{\partial x_1}, \ldots, \frac{\partial f(x)}{\partial x_n} \right]^T ,$$

(recall our convention that all vectors are column vectors). Given a scalar-valued function $u : \prod_{i=1}^{I} \mathbb{R}^{m_i} \to \mathbb{R}$, we use the notation $\nabla_i u(x)$ to denote the gradient vector of u with respect to x_i at point x, i.e.,

$$\nabla_i u(x) = \left[\frac{\partial u(x)}{\partial x_i^1}, \ldots, \frac{\partial u(x)}{\partial x_i^{m_i}} \right]^T . \tag{1.15}$$

We next state necessary conditions for the optimality of a feasible solution of a nonlinear optimization problem. These conditions are referred to as the *Karush-Kuhn-Tucker conditions* in the optimization literature.

Theorem 1.45   **(Karush-Kuhn-Tucker conditions)** *Let x^* be an optimal solution of the optimization problem*

$$\begin{aligned} \text{maximize} \quad & f(x) \\ \text{subject to} \quad & g_j(x) \geq 0, \qquad j = 1, \ldots, r, \end{aligned}$$

where the cost function $f : \mathbb{R}^n \to \mathbb{R}$ and the constraint functions $g_j : \mathbb{R}^n \to \mathbb{R}$ are continuously differentiable. Denote the set of active constraints at x^* as $A(x^*) = \{j = 1, \ldots, r \mid g_j(x^*) = 0\}$. Assume that the active constraint gradients, $\nabla g_j(x^*)$, $j \in A(x^*)$, are linearly independent vectors. Then, there exists a nonnegative vector $\lambda^* \in \mathbb{R}^r$ (Lagrange multiplier vector) such that

$$\nabla f(x^*) + \sum_{j=1}^{r} \lambda_j^* \nabla g_j(x^*) = 0,$$

$$\lambda_j^* g_j(x^*) = 0, \qquad \forall \, j = 1, \ldots, r. \tag{1.16}$$

One can view the preceding theorem as a generalization of the Lagrange multiplier theorem for equality constrained optimization problems to inequality constrained optimization problems. The condition that the active constraint gradients, $\nabla g_j(x^*)$, $j \in A(x^*)$, are linearly independent vectors is a regularity condition that eliminates "degenerate" cases where there may not exist Lagrange multipliers. These type of conditions are referred to as *constraint qualifications*. The condition in Eq. (1.16) implies that if $\lambda_j^* > 0$, then $g_j(x^*) = 0$, and if $g_j(x^*) > 0$, then $\lambda_j^* = 0$. It is therefore referred to as the *complementary slackness condition* and captures loosely the intuitive idea that a multiplier is used only when the constraint is active and the problem is locally unconstrained with respect to the inactive constraints.

For convex optimization problems (i.e., minimizing a convex function over a convex constraint set, or maximizing a concave function over a convex constraint set), we can provide necessary and sufficient conditions for the optimality of a feasible solution:

Theorem 1.46   *Consider the optimization problem*

$$\begin{aligned} maximize \quad & f(x) \\ subject\ to \quad & g_j(x) \geq 0, \qquad j = 1, \ldots, r, \end{aligned}$$

where the cost function $f : \mathbb{R}^n \to \mathbb{R}$ and the constraint functions $g_j : \mathbb{R}^n \to \mathbb{R}$ are concave functions. Assume also that there exists some $\bar{x}$ such that $g_j(\bar{x}) > 0$ for all $j = 1, \ldots, r$. Then a vector $x^* \in \mathbb{R}^n$ is an optimal solution of the preceding problem if and only if $g_j(x^*) \geq 0$ for all $j = 1, \ldots, r$, and there exists a nonnegative vector $\lambda^* \in \mathbb{R}^r$ (Lagrange multiplier vector) such that

$$\nabla f(x^*) + \sum_{j=1}^{r} \lambda_j^* \nabla g_j(x^*) = 0,$$

$$\lambda_j^* g_j(x^*) = 0, \qquad \forall\, j = 1, \ldots, r.$$

Note that the condition that there exists some $\bar{x}$ such that $g_j(\bar{x}) > 0$ for all $j = 1, \ldots, r$ is a constraint qualification (referred to as the *Slater's constraint qualification*) that can be used for convex optimization problems.

CHAPTER 2

Game Theory Dynamics

In the previous chapter, we have studied strategic form games which are used to model static games where each player chooses his action once and for all, simultaneously. This chapter presents two complementary ways for introducing dynamic analysis in competitive situations. First, we study extensive form (dynamic) games, which are used to model dynamic decision making in multi-agent environments. Second, we analyze dynamics induced by the repeated play of the same strategic form game by boundedly rational agents, following simple myopic rules.

The last part of the chapter studies special classes of games with appealing dynamic properties, supermodular and potential games, which have widespread applications for network game models.

2.1 EXTENSIVE FORM GAMES

In this section, we present extensive form games which model multi-agent sequential decision making. Our focus will be on *multi-stage games with observed actions* where: (1) all previous actions are observed, i.e., each player is perfectly informed of all previous events; (2) some players may move simultaneously at some stage k. Extensive form games can be conveniently represented by *game trees* as illustrated in the next examples.

Example 2.1

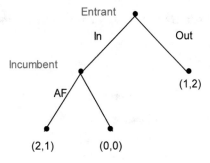

Figure 2.1: Entry deterrence game.

We first consider an entry deterrence game, in which there are two players (see Figure 2.1). Player 1, the entrant, can choose to enter the market or stay out. Player 2, the incumbent, after observing the action of the entrant, chooses to accommodate him or fight with him. The payoffs for each of the action profiles (or histories) are given by the pair (x, y) at the leaves of the game tree: x denotes the payoff of player 1 (the entrant) and y denotes the payoff of player 2 (the incumbent).

Example 2.2

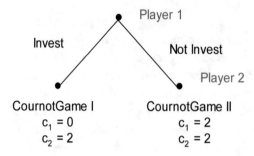

Figure 2.2: Investment in duopoly.

Our next example considers a duopoly investment game, in which there are two players in the market (see Figure 2.2). Player 1 can choose to invest or not invest. After player 1 chooses his action, both players engage in a Cournot competition (cf. Example 1.3). If player 1 invests, then they will engage in a Cournot game with $c_1 = 0$ and $c_2 = 2$. Otherwise, they will engage in a Cournot game with $c_1 = c_2 = 2$. We can also assume that there is a fixed cost of f for player 1 to invest.

We next formally define the extensive form game model.

Definition 2.3   Extensive Form Game   An extensive form game G consists of the following components:

1. A set of players, $\mathcal{I} = \{1, \ldots, I\}$.

2. A set H of sequences, referred to as *histories*, defined as follows:

$$\begin{aligned}
h^0 &= \emptyset & &\text{initial history} \\
s^0 &= (s_1^0, \ldots, s_I^0) & &\text{stage 0 action profile} \\
h^1 &= s^0 & &\text{history after stage 0}
\end{aligned}$$

$$\vdots \qquad\qquad\qquad \vdots$$

$$h^{k+1} = (s^0, s^1, \ldots, s^k) \quad \text{history after stage } k$$

Let $H^k = \{h^k\}$ be the set of all possible stage k histories. Then, $H = \cup_{k=0}^{\infty} H^k$ is the set of all possible histories. If the game has a finite number $(K + 1)$ of stages, then it is a *finite horizon game*. We use H^{K+1} to denote the set of all possible *terminal histories*.

3. A set of pure strategies for each player, defined as a contingency plan of how to play in each stage k for every possible history h^k. Let $S_i(H^k) = \bigcup_{h^k \in H^k} S_i(h^k)$ be the set of actions available to player i at stage k. Define $s_i^k : H^k \to S_i(H^k)$ such that $s_i(h^k) \in S_i(h^k)$. Then the pure strategy of player i is the set of sequences $s_i = \{s_i^k\}_{k=0}^{\infty}$, i.e., a pure strategy of a player is a collection of functions from all possible histories into available actions. A strategy profile s is given by the tuple $s = (s_1, \dots, s_I)$. Given a strategy profile s, we can find the sequence of actions generated: the stage 0 actions are $s^0 = s^0(h^0)$, the stage 1 actions are $s^1 = s^1(s^0)$, the stage 2 actions are $s^2 = s^2(s^0, s^1)$, and so on. This is called the *path or outcome of strategy profile s*.

4. A set of preferences for each player. Since for finite horizon games, terminal histories H^{K+1} specify an entire sequence of play, we can represent the preferences of player i by a utility function $u_i : H^{K+1} \to \mathbb{R}$. In most applications, the utility functions are additively separable over stages, i.e., each player's utility is some weighted average of single stage payoffs. As the strategy profile s determines the path $s^0, \dots, s^k$ and hence h^{K+1}, we will use the notation $u_i(s)$ as the payoff to player i under the strategy profile s.

The next example illustrates pure strategies of players in extensive form games.

Example 2.4

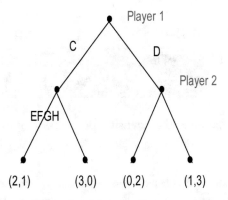

Figure 2.3: Strategies in an extensive form game.

We consider the extensive form game illustrated in Figure 2.3. Player 1's strategies are given by functions $s_1 : H^0 = \emptyset \to S_1 = \{C, D\}$, which can be represented as two possible strategies; C,D. Player 2's strategies are given by functions $s^2 : H^1 = \{\{C\}, \{D\}\} \to S_2$, which can be represented as four possible strategies; EG, EH, FG and FH. For the strategy profile $s = (C, EG)$, the outcome is given by $\{C, E\}$. Similarly, for the strategy profile $s = (D, EG)$, the outcome will be $\{D, G\}$.

In the next example, we determine the pure strategies of players in an extensive form game and use it to define the *strategic (or normal) form representation of an extensive form game.*

Example 2.5   Consider the following two-stage extensive form version of matching pennies.

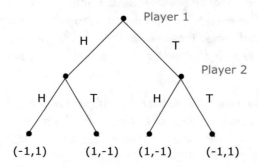

Figure 2.4: Two-stage extensive form version of matching pennies.

Since a strategy should be a complete contingency plan, player 2 has 4 different pure strategies: heads following heads, heads following tails HH; heads following heads, tails following tails HT; tails following heads, tails following tails TT; tails following heads, heads following tails TH. Identifying the pure strategies of players allows us to go from the extensive form game to its strategic form representation:

	HH	HT	TT	TH
H	$-1, 1$	$-1, 1$	$1, -1$	$1, -1$
T	$1, -1$	$-1, 1$	$-1, 1$	$1, -1$

Strategic form representation of the extensive form version of matching pennies given in Figure 2.4.

A Nash equilibrium of an extensive form game can be defined through its strategic form representation, i.e., a pure strategy Nash equilibrium of an extensive form game is a strategy profile s such that no player i can do better with a different strategy, which is the familiar condition that $u_i(s_i, s_{-i}) \geq u_i(s_i', s_{-i})$ for all s_i'.[1] The next example finds the Nash equilibria of an extensive form game and argues that not all Nash equilibria represent reasonable outcomes in extensive form games.

Example 2.6   We consider the entry deterrence game described in Example 2.1. The equivalent strategic form representation of this game is given by:

[1]A mixed strategy Nash equilibrium is defined similarly.

	ACCOMMODATE	FIGHT
IN	(2, 1)	(0, 0)
OUT	(1, 2)	(1, 2)

Strategic form representation of the entry deterrence game illustrated in Figure 2.1.

Hence, this game has two pure Nash equilibria: the strategies (IN, ACCOMMODATE) and (OUT, FIGHT). However, the equilibrium (OUT, FIGHT) does not seem reasonable: the entrant chooses "Out" because he believes that the incumbent will fight if he enters the market. However, if this happens, it is more profitable for the incumbent to choose "Accommodate". Hence, the equilibrium (OUT, FIGHT) is sustained by a *noncredible threat* of the incumbent.

This motivates a new equilibrium notion for extensive form games, *Subgame Perfect (Nash) Equilibrium*, which requires each player's strategy to be "optimal," not only at the start of the game, but also after every history. To define subgame perfection formally, we first define the notion of a subgame. In perfect information extensive form games (i.e., each player is perfectly informed about all the actions chosen at the previous stages when choosing an action), each node of the game tree corresponds to a unique history, and is referred to as a *subgame*.[2] We use the notation h^k or $G(h^k)$ to denote a subgame. A restriction of a strategy s to subgame G', $s_{|G'}$ is the action profile implied by s in the subgame G'.

Definition 2.7 Subgame Perfect Equilibrium   A strategy profile s^* is a *Subgame Perfect Nash equilibrium* (SPE) in game G if for any subgame G' of G, $s^*_{|G'}$ is a Nash equilibrium of G'.

Subgame perfection (i.e., the fact that a strategy profile restricted to any subgame should be a Nash equilibrium of the subgame) will remove noncredible threats since these will not be Nash equilibria in the appropriate subgames. In the entry deterrence game, following entry, the action "Fight" is not a best response, and thus not a Nash equilibrium of the corresponding subgame. Therefore, (Out, Fight) is not an SPE.

To find the SPE of an extensive form game, one can find all the Nash equilibria and eliminate those that are not subgame perfect. However, for finite horizon games, the SPE can be found more economically by using *backward induction*. Backward induction refers to starting from the last subgames of a finite game, then finding the best response strategy profiles or the Nash equilibria in the subgames, then assigning these strategies profiles and the associated payoffs to the subgames, and moving successively towards the beginning of the game. It is straightforward to see that backward induction provides the entire set of SPE.

---

[2]In imperfect information extensive form games, subgames are defined through *information sets*, which model the information that players have when they are choosing their actions. Information sets partition the nodes of a game tree and can be viewed as a generalization of a history; see [46].

For general extensive form games (i.e., finite or infinite horizon games), we will rely on a useful characterization of the subgame perfect equilibria given by the "one stage deviation principle," which is essentially the *principle of optimality* of dynamic programming. We first state it for finite horizon games.

Theorem 2.8   One stage deviation principle   *For finite horizon multi-stage games with observed actions, s^* is a subgame perfect equilibrium if and only if for all i, t and h^t, we have*

$$u_i(s_i^*, s_{-i}^*|h^t) \geq u_i(s_i, s_{-i}^*|h^t)$$

for all s_i satisfying

$$
\begin{aligned}
s_i(h^t) &\neq s_i^*(h^t), \\
s_{i|h^t}(h^{t+k}) &= s_{i|h^t}^*(h^{t+k}), \quad \text{for all } k > 0, \text{ and all } h^{t+k} \in G(h^t).
\end{aligned}
$$

Informally, a strategy profile s is an SPE if and only if no player i can gain by deviating from s in a single stage and conforming to s thereafter. We omit the proof of the one stage deviation principle for finite horizon games, which uses a recursive argument to show that if a strategy satisfies the one stage deviation principle, then that strategy cannot be improved upon by a finite number of deviations. This leaves open the possibility that a player may gain by an infinite sequence of deviations, which we exclude using the following condition.

Definition 2.9   Consider an extensive form game with an infinite horizon, denoted by G^∞. Let h denote an ∞-horizon history, i.e., $h = (s^0, s^1, s^2...)$, is an infinite sequence of actions. Let $h^t = (s^0, \ldots, s^{t-1})$ be the restriction to first t periods. The game G^∞ is *continuous at infinity* if for all players i, the payoff function u_i satisfies

$$\sup_{i,h,\tilde{h} \text{ s.t. } h^t = \tilde{h}^t} |u_i(h) - u_i(\tilde{h})| \to 0 \quad \text{as} \quad t \to \infty.$$

The continuity at infinity condition is satisfied when the overall payoffs are a discounted sum of stage payoffs, i.e.,

$$u_i = \sum_{t=0}^{\infty} \delta_i^t g_i^t(s^t),$$

(where $g_i^t(s^t)$ are the stage payoffs, and the positive scalar $\delta_i < 1$ is a discount factor), and the stage payoff functions are uniformly bounded, i.e., there exists some B such that $\max_{t,s^t} |g_i^t(s^t)| < B$.

Theorem 2.10   *Consider an infinite-horizon game, G^∞, that is continuous at infinity. Then, the one stage deviation principle holds, i.e., the strategy profile s^* is an SPE if and only if for all i, h^t, and t, we have*

$$u_i(s_i^*, s_{-i}^*|h^t) \leq u_i(s_i, s_{-i}^*|h^t),$$

for all s_i that satisfies $s_i(h^t) \neq s_i^*(h^t)$ and $s_{i|h^t}(h^{t+k}) = s_{i|h^t}^*(h^{t+k})$ for all $h^{t+k} \in G(h^t)$ and for all $k > 0$.

2.2   LEARNING DYNAMICS IN GAMES – FICTITIOUS PLAY

Most economic theory relies on equilibrium analysis, based on Nash equilibrium or its refinements. The traditional explanation for when and why equilibrium arises is that it results from analysis and introspection by the players in a situation where the rules of the game, the rationality of the players, and the payoff functions of players are all common knowledge.

In this section, we develop an alternative explanation why equilibrium arises using a dynamic process in which less than fully rational players grope for optimality over time. For this, we rely on adaptive models studied in learning in games literature which prescribe rules for how players form beliefs and select strategies over time (see [45]). Our focus will be on *fictitious play* (and its variants), introduced by Brown [28], which is a widely used model of learning. In this process, agents behave assuming they are facing a stationary, but unknown, distribution of opponents' strategies. We examine whether fictitious play is a sensible model of learning and the asymptotic behavior of play when all players use fictitious play learning rules.

The most compelling justification of fictitious play is as a "belief-based" learning rule, i.e., players form beliefs about opponent play (from the entire history of past play) and behave rationally with respect to these beliefs.

The model for fictitious play is given below. Here we focus on the case where we have two players only, see [45] for the analysis of fictitious play for multiple players.

- We have two players playing the strategic form game G at times $t = 1, 2, \ldots$.

- We denote the stage payoff of player i by $g_i(s_i, s_{-i})$.

- For $t = 1, 2, \ldots$ and $i = 1, 2$, we define the function $\eta_i^t : S_{-i} \to \mathbb{R}$, where $\eta_i^t(s_{-i})$ is the number of times player i has observed the action s_{-i} before time t. We use $\eta_i^0(s_{-i})$ to represent a starting point (or fictitious past).

Example 2.11   Assume that $S_2 = \{U, D\}$. If $\eta_1^0(U) = 3$ and $\eta_1^0(D) = 5$, and player 2 plays U, U, D in the first three periods, then $\eta_1^3(U) = 5$ and $\eta_1^3(D) = 6$.

Each player assumes that his opponent is using a stationary mixed strategy. Players choose actions in each period (or stage) to maximize that period's expected payoff given their prediction of the distribution of opponent's actions, which they form according to:

$$\mu_i^t(s_{-i}) = \frac{\eta_i^t(s_{-i})}{\sum_{\bar{s}_{-i} \in S_{-i}} \eta_i^t(\bar{s}_{-i})},$$

i.e., player i forecasts player $-i$'s strategy at time t to be the empirical frequency distribution of past play.

Given player i's belief or forecast about his opponents play, he chooses his action at time t to maximize his payoff, i.e.,

$$s_i^t \in \arg\max_{s_i \in S_i} g_i(s_i, \mu_i^t).$$

This choice is *myopic*, i.e., players are trying to maximize current payoff without considering their future payoffs. Note that the myopic assumption is consistent with the assumption that players are using stationary mixed strategies.

Example 2.12   Consider the fictitious play of the following game:

	L	R
U	3, 3	0, 0
D	4, 0	1, 1

Note that this game is dominance solvable (D is a strictly dominant strategy for the row player), and the unique Nash equilibrium is (D, R). Assume that $\eta_1^0 = (3, 0)$ and $\eta_2^0 = (1, 2.5)$.
Period 1: Then, $\mu_1^0 = (1, 0)$ and $\mu_2^0 = (1/3.5, 2.5/3.5)$, so play follows $s_1^0 = D$ and $s_2^0 = L$.
Period 2: We have $\eta_1^1 = (4, 0)$ and $\eta_2^1 = (1, 3.5)$, so play follows $s_1^1 = D$ and $s_2^1 = R$.
Period 3: We have $\eta_1^1 = (4, 1)$ and $\eta_2^1 = (1, 4.5)$, so play follows $s_1^2 = D$ and $s_2^2 = R$.
Since D is a dominant strategy for the row player, he always plays D, and μ_2^t converges to $(0, 1)$ with probability 1. Therefore, player 2 will end up playing R.

The striking feature of the fictitious play is that players do not have to know anything about their opponent's payoff. They only form beliefs about how their opponents will play.

2.2.1   CONVERGENCE OF FICTITIOUS PLAY

Let $\{s^t\}$ be a sequence of strategy profiles generated by fictitious play. In this section, we study the asymptotic behavior of the sequence $\{s^t\}$, i.e., the convergence properties of the sequence $\{s^t\}$ as t goes to infinity.

We first define the notion of convergence to pure strategies.

Definition 2.13   The sequence $\{s^t\}$ converges to s if there exists T such that $s^t = s$ for all $t \geq T$.

The next proposition formalizes the property that if the fictitious play sequence converges, then it must converge to a Nash equilibrium of the game.

Proposition 2.14   *Let $\{s^t\}$ be a sequence of strategy profiles generated by fictitious play.*

(a) *If $\{s^t\}$ converges to $\bar{s}$, then $\bar{s}$ is a pure strategy Nash equilibrium.*

(b) Suppose that for some t, $s^t = s^*$, where s^* is a strict Nash equilibrium of G. Then $s^\tau = s^*$ for all $\tau > t$.

The proof of part (a) is straightforward. We provide a proof for part (b):

Proof of part (b): Let $s^t = s^*$. We will show that $s^{t+1} = s^*$. Note that

$$\mu_i^{t+1} = (1 - \alpha)\mu_i^t + \alpha s_{-i}^t = (1 - \alpha)\mu_i^t + \alpha s_{-i}^*,$$

where, abusing the notation, we used s_{-i}^t to denote the degenerate probability distribution and

$$\alpha = \frac{1}{\sum_{s_{-i}} \eta_i^t(s_{-i}) + 1}.$$

Therefore, by the linearity of the expectation, we have for all $s_i \in S_i$,

$$g_i(s_i, \mu_i^{t+1}) = (1 - \alpha)g_i(s_i, \mu_i^t) + \alpha g_i(s_i, s_{-i}^*).$$

Since s_i^* maximizes both terms (in view of the fact that s^* is a strict Nash equilibrium), it follows that s_i^* will be played at $t + 1$. □

Note that the preceding notion of convergence only applies to pure strategies. We next provide an alternative notion of convergence, i.e., convergence of empirical distributions or beliefs.

Definition 2.15   The sequence $\{s^t\}$ *converges to* $\sigma \in \Sigma$ *in the time-average sense* if for all i and for all $s_i \in S_i$, we have

$$\lim_{T \to \infty} \frac{[\text{number of times } s_i^t = s_i \text{ for } t \leq T]}{T + 1} = \sigma(s_i),$$

i.e., $\mu_{-i}^T(s_i)$ converges to $\sigma_i(s_i)$ as $T \to \infty$.

The next example illustrates convergence of the fictitious play sequence in the time-average sense.

Example 2.16   Matching Pennies   Consider the fictitious play of the following matching pennies game:

	H	T
H	$1, -1$	$-1, 1$
T	$-1, 1$	$1, -1$

Consider the following sequence of play:

Time	η_1^t	η_2^t	Play
0	$(0,0)$	$(0,2)$	(H,H)
1	$(1,0)$	$(1,2)$	(H,H)
2	$(2,0)$	$(2,2)$	(H,T)
3	$(2,1)$	$(3,2)$	(H,T)
4	$(2,2)$	$(4,2)$	(T,T)
5	$(2,3)$	$(4,3)$	(T,T)
6	...	...	(T,H)

Play continues as (T,H), (H,H), (H,H) - a deterministic cycle. The time average converges to $\big((1/2, 1/2), (1/2, 1/2)\big)$, which is the unique Nash equilibrium.

The next proposition extends the basic convergence property of fictitious play, stated in Proposition 2.14, to mixed strategies.

Proposition 2.17   *Suppose a fictitious play sequence $\{s^t\}$ converges to σ in the time-average sense. Then σ is a Nash equilibrium of G.*

Proof. Suppose s^t converges to σ in the time-average sense. Assume that σ is not a Nash equilibrium. Then there exist some $i, s_i, s_i' \in S_i$ with $\sigma_i(s_i) > 0$ such that

$$g_i(s_i', \sigma_{-i}) > g_i(s_i, \sigma_{-i}).$$

Choose $\epsilon > 0$ such that

$$\epsilon < \frac{1}{2}\Big[g_i(s_i', \sigma_{-i}) - g_i(s_i, \sigma_{-i})\Big],$$

and T sufficiently large that for all $t \geq T$, we have

$$\left|\mu_i^T(s_{-i}) - \sigma_{-i}(s_{-i})\right| < \frac{\epsilon}{\max_{s \in S} g_i(s)},$$

which is possible since $\mu_i^t \to \sigma_{-i}$ by assumption. Then, for any $t \geq T$, we have

$$
\begin{aligned}
g_i(s_i, \mu_i^t) &= \sum_{s_{-i}} g_i(s_i, s_{-i})\mu_i^t(s_{-i}) \\
&\leq \sum_{s_{-i}} g_i(s_i, s_{-i})\sigma_{-i}(s_{-i}) + \epsilon \\
&< \sum_{s_{-i}} g_i(s_i', s_{-i})\sigma_{-i}(s_{-i}) - \epsilon \\
&\leq \sum_{s_{-i}} g_i(s_i', s_{-i})\mu_i^t(s_{-i}) = g_i(s_i', \mu_i^t).
\end{aligned}
$$

This shows that after T, s_i is never played, implying that as $T \to \infty$, $\mu_{-i}^t(s_i) \to 0$. But this contradicts the fact that $\sigma_i(s_i) > 0$, completing the proof. $\qquad\square$

It is important to realize that convergence in the time-average sense is not necessarily a natural convergence notion, as illustrated in the following example.

Example 2.18 Mis-coordination   Consider the fictitious play of the following game:

	A	B
A	1, 1	0, 0
B	0, 0	1, 1

Note that this game had a unique mixed Nash equilibrium $\big((1/2, 1/2), (1/2, 1/2)\big)$. Consider the following sequence of play:

Time	η_1^t	η_2^t	Play
0	$(1/2, 0)$	$(0, 1/2)$	(A, B)
1	$(1/2, 1)$	$(1, 1/2)$	(B, A)
2	$(3/2, 1)$	$(1, 3/2)$	(A, B)
3	...	...	(B, A)
4	...	...	(A, B)

Play continues as (A,B), (B,A), . . . - again a deterministic cycle. The time average converges to $\big((1/2, 1/2), (1/2, 1/2)\big)$, which is a mixed strategy equilibrium of the game. But players never successfully coordinate!

The following proposition provides some classes of games for which fictitious play converges in the time-average sense (see Fudenberg and Levine [45]; see also [87], [72], [75] for the proofs).

Proposition 2.19   *Fictitious play converges in the time-average sense for the game G under any of the following conditions:*

(a) G is a two player zero-sum game.

(b) G is a two player nonzero-sum game where each player has at most two strategies.

(c) G is solvable by iterated strict dominance.

(d) G is an identical interest game, i.e., all players have the same payoff function.

(e) G is an (exact) potential game (see Def. 2.36).

In Section 2.2.3, we consider a continuous time variant of fictitious play and provide the proof of convergence for two player zero-sum games and identical interest games.

2.2.2 NON-CONVERGENCE OF FICTITIOUS PLAY

The next proposition shows that there exists games in which fictitious play does not converge in the time-average sense.

Proposition 2.20 Shapley [94] *Fictitious play does not converge in Shapley's modified version of Rock–Scissors–Paper game (described below).*

Example 2.21 Shapley's modified version of Rock-Scissors-Paper game has payoffs:

	R	S	P
R	0,0	1,0	0,1
S	0,1	0,0	1,0
P	1,0	0,1	0,0

Suppose that $\eta_1^0 = (1, 0, 0)$ and that $\eta_2^0 = (0, 1, 0)$. Then in period 0, play is (P,R). In period 1, player 1 expects R, and 2 expects S, so play is (P,R). Play then continues to follow (P,R) until player 2 switches to S (suppose this lasts for k periods). Play then follows (P,S), until player 1 switches to R (for βk periods, where β is a scalar such that $\beta > 1$). Play then follows (R,S), until player 2 switches to P (for $\beta^2 k$ periods). It can be shown that play of the game cycles among 6 (off-diagonal) profiles with periods of ever-increasing length, thus non-convergence.

Shapley's proof of the non-convergence of fictitious play in the above game explicitly computes time spent in each stage of the sequence. Here, we provide an alternative insightful proof due to Monderer *et al.* [74]. For this we will make use of the following lemma, which relates realized payoffs to expected payoffs under the empirical distributions. Define the time-average payoffs through time t as:

$$U_i^t = \frac{1}{t+1} \sum_{\tau=0}^{t} g_i(s_i^\tau, s_{-i}^\tau).$$

Define the expected payoffs at time t as:

$$\tilde{U}_i^t = g_i(s_i^t, \mu_i^t) = \max_{s_i \in S_i} g_i(s_i, \mu_i^t).$$

Lemma 2.22 *Given any $\epsilon > 0$ and any $i \in \mathcal{I}$, there exists some T such that for all $t \geq T$, we have*

$$\tilde{U}_i^t \geq U_i^t - \epsilon.$$

Proof. Note that

$$
\begin{aligned}
\tilde{U}_i^t = g_i(s_i^t, \mu_i^t) &\geq g_i(s_i^{t-1}, \mu_i^t) \\
&= \frac{1}{t+1} g_i(s_i^{t-1}, s_{-i}^{t-1}) + \frac{t}{t+1} g_i(s_i^{t-1}, \mu_i^{t-1}) \\
&= \frac{1}{t+1} g_i(s_i^{t-1}, s_{-i}^{t-1}) + \frac{t}{t+1} \tilde{U}_i^{t-1}.
\end{aligned}
$$

Expanding $\tilde{U}_i^{t-1}$ and iterating, we obtain

$$
\tilde{U}_i^t \geq \frac{1}{t+1} \sum_{\tau=0}^{t} g_i(s_i^\tau, s_{-i}^\tau) - \frac{1}{t+1} g_i(s_i^t, s_{-i}^t).
$$

Choosing T such that

$$
\epsilon > \frac{1}{T+1} \max_{s \in S} g_i(s),
$$

completes the proof. $\square$

We can now use the preceding lemma to prove Proposition 2.20 (Shapley's result).

Proof of Proposition 2.20. In the Rock-Scissors-Paper game, there is a unique Nash equilibrium with expected payoffs of 1/3 for both players. Therefore, if the fictitious play converged, then $\tilde{U}_i^t \to 1/3$ for $i = 1, 2$, implying that $\tilde{U}_1^t + \tilde{U}_2^t \to 2/3$. But under fictitious play, realized payoffs always sum to 1, i.e.,

$$
U_1^t + U_2^t = 1, \qquad \text{for all } t,
$$

thus contradicting the lemma above, and showing that fictitious play does not converge. $\square$

2.2.3 CONVERGENCE PROOFS

We now consider a continuous time version of fictitious play and provide a proof of convergence for zero-sum games and identical interest games.

2.2.3.1 Continuous Time Fictitious Play

Let us introduce some notation to facilitate the subsequent analysis. We denote the empirical distribution of player i's play up to (but not including) time t when time intervals are of length Δt by

$$
p_i^t(s_i) = \frac{\sum_{\tau=0}^{(t-\Delta t)/\Delta t} \mathcal{I}\{s_i^\tau = s_i\}}{t/\Delta t}.
$$

We use $p^t \in \Sigma$ to denote the product distribution formed by the p_i^t. We can now think of making time intervals Δt smaller, which will lead us to a version a fictitious play in continuous time.

In *continuous time fictitious play* (CTFP), the empirical distributions of the players are updated in the direction of a best response to their opponents' past action:

$$\frac{dp_i^t}{dt} \in BR_i(p_{-i}^t) - p_i^t,$$

where

$$BR_i(p_{-i}^t) = \arg\max_{\sigma_i \in \Sigma_i} g_i(\sigma_i, p_{-i}^t).$$

Another variant of the CTFP is the *perturbed CTFP* defined by

$$\frac{dp_i^t}{dt} = C_i(p_{-i}^t) - p_i^t,$$

where

$$C_i(p_{-i}^t) = \arg\max_{\sigma_i \in \Sigma_i} \left[g_i(\sigma_i, p_{-i}^t) - V_i(\sigma_i) \right], \tag{2.1}$$

and $V_i : \Sigma_i \to \mathbb{R}$ is a strictly convex function and satisfies a "boundary condition" (see Fudenberg and Levine [45] for examples of perturbation functions). Note that because the perturbed best-response C_i is uniquely defined, the perturbed CTFP is described by a differential equation rather than a differential inclusion.

2.2.3.2   Convergence of (perturbed) CTFP for Zero-sum Games

In this section, we provide a proof of convergence of perturbed CTFP for two player zero-sum games.

We consider a two player zero-sum game with payoff matrix M, where the payoffs are perturbed by the functions V_i (defined above), i.e., the payoffs are given by

$$\Pi_1(\sigma_1, \sigma_2) = \sigma_1' M \sigma_2 - V_1(\sigma_1), \tag{2.2}$$

$$\Pi_2(\sigma_1, \sigma_2) = -\sigma_1' M \sigma_2 - V_2(\sigma_2). \tag{2.3}$$

Let $\{p^t\}$ be generated by the perturbed CTFP,

$$\frac{dp_i^t}{dt} = C_i(p_{-i}^t) - p_i^t,$$

[cf. Eq. (2.1)]. We use a Lyapunov function approach to prove convergence. In particular, we consider the function

$$W(t) = U_1(p^t) + U_2(p^t),$$

where the functions $U_i : \Sigma \to \mathbb{R}$ are defined as

$$U_i(\sigma_i, \sigma_{-i}) = \max_{\sigma_i' \in \Sigma_i} \Pi_i(\sigma_i', \sigma_{-i}) - \Pi_i(\sigma_i, \sigma_{-i}),$$

(i.e., the function U_i gives the maximum possible payoff improvement player i can achieve by a unilateral deviation in his own mixed strategy). Note that $U_i(\sigma) \geq 0$ for all $\sigma \in \Sigma$, and $U_i(\sigma) = 0$ for all i implies that σ is a mixed Nash equilibrium.

For the zero sum game with payoffs (2.2)–(2.3), the function $W(t)$ takes the form

$$W(t) = \max_{\sigma_1' \in \Sigma_1} \Pi_1(\sigma_1', p_2^t) + \max_{\sigma_2' \in \Sigma_2} \Pi_i(p_1^t, \sigma_2') + V_1(p_1^t) + V_2(p_2^t).$$

We will show that $\frac{dW(t)}{dt} \leq 0$ with equality if and only if $p_i^t = C_i(p_{-i}^t)$, showing that for all initial conditions p^0, we have

$$\lim_{t \to \infty} \left(p_i^t - C_i(p_{-i}^t) \right) = 0 \qquad i = 1, 2.$$

We need the following lemma.

Lemma 2.23   **(Envelope Theorem)** *Let $F : \mathbb{R}^n \times \mathbb{R}^m \to \mathbb{R}$ be a continuously differentiable function. Let $U \subset \mathbb{R}^m$ be an open convex subset, and $u^*(x)$ be a continuously differentiable function such that*

$$F(x, u^*(x)) = \min_{u \in U} F(x, u).$$

Let $H(x) = \min_{u \in U} F(x, u)$. Then,

$$\nabla_x H(x) = \nabla_x F(x, u^*(x)).$$

Proof. The gradient of $H(x)$ is given by

$$\begin{aligned} \nabla_x H(x) &= \nabla_x F(x, u^*(x)) + \nabla_u F(x, u^*(x)) \nabla_x u^*(x) \\ &= \nabla_x F(x, u^*(x)), \end{aligned}$$

where we use the fact that $\nabla_u F(x, u^*(x)) = 0$ since $u^*(x)$ minimizes $F(x, u)$.   □

Using the preceding lemma, we have

$$\begin{aligned} \frac{d}{dt} \left[\max_{\sigma_1' \in \Sigma_1} \Pi_1(\sigma_1', p_2^t) \right] &= \nabla_{\sigma_2} \Pi_1(C_1(p_2^t), p_2^t)' \frac{dp_2^t}{dt} \\ &= C_1(p_2^t)' M \frac{dp_2^t}{dt} \\ &= C_1(p_2^t)' M \left(C_2(p_1^t) - p_2^t \right). \end{aligned}$$

Similarly,

$$\frac{d}{dt} \left[\max_{\sigma_2' \in \Sigma_2} \Pi_2(p_1^t, \sigma_2') \right] = -(C_1(p_2^t) - p_1^t)' M C_2(p_1^t).$$

Combining the preceding two relations, we obtain

$$\frac{dW(t)}{dt} = -C_1(p_2^t)'Mp_2^t + (p_1^t)'MC_2(p_1^t) + \nabla V_1(p_1^t)'\frac{dp_1^t}{dt} + \nabla V_2(p_2^t)'\frac{dp_2^t}{dt}. \qquad (2.4)$$

Since $C_i(p_{-i}^t)$ is a perturbed best response, we have

$$C_1(p_2^t)'Mp_2^t - V_1(C_1(p_2^t)) \geq (p_1^t)'Mp_2^t - V_1(p_1^t),$$

$$-(p_1^t)'MC_2(p_1^t) - V_2(C_2(p_1^t)) \geq -(p_1^t)'Mp_2^t - V_2(p_2^t),$$

with equality if and only if $C_i(p_{-i}^t) = p_i^t, i = 1, 2$ (the latter claim follows by the uniqueness of the perturbed best response). Combining these relations, we have

$$
\begin{aligned}
-C_1(p_2^t)'Mp_2^t + (p_1^t)'MC_2(p_1^t) &\leq \sum_i [V_i(p_i^t) - V_i(C_i(p_{-i}^t))] \\
&\leq \sum_i \nabla V_i(p_i^t)'(C_i(p_i^t) - p_i^t) \\
&= -\sum_i \nabla V_i(p_i^t)'\frac{dp_i^t}{dt},
\end{aligned}
$$

where the second inequality follows by the convexity of V_i. The preceding relation and Eq. (2.4) imply that $\frac{dW}{dt} \leq 0$ for all t, with equality if and only if $C_i(p_{-i}^t) = p_i^t$ for both players, completing the proof.

2.2.3.3   Convergence of CTFP for Identical Interest Games

We finally provide a proof of convergence of CTFP for identical interest games. Consider an I-player game with *identical interests*, i.e., a game where all players share the same payoff function Π. Recall the continuous time fictitious play (CTFP) dynamics:

$$\frac{dp_i^t}{dt} \in BR_i(p_{-i}^t) - p_i^t.$$

Let $\{p_i^t\}$ denote the sequence generated by the CTFP dynamics and let $\sigma_i^t = p_i^t + dp_i^t/dt$. Note that $\sigma_i^t \in BR_i(p_{-i}^t)$.

Theorem 2.24   *For all players i and regardless of the initial condition p^0, we have*

$$\lim_{t\to\infty} \left[\max_{\sigma_i'\in\Sigma_i} \Pi(\sigma_i', p_{-i}^t) - \Pi(p_i^t, p_{-i}^t) \right] = 0;$$

namely, p_i^t is asymptotically a best response to p_{-i}^t.

Proof. We again consider the function $W(t) \equiv \sum_i U_i(p^t)$, where

$$U_i(\sigma_i, \sigma_{-i}) = \max_{\sigma_i' \in \Sigma} \Pi(\sigma_i', \sigma_{-i}) - \Pi(\sigma_i, \sigma_{-i}).$$

Observe that

$$
\begin{aligned}
\frac{d}{dt}(\Pi(p^t)) &= \frac{d}{dt}\left[\sum_{s_i \in S_i} \cdots \sum_{s_n \in S_n} p_1^t(s_1) \cdots p_n^t(s_n)\Pi(s)\right] \\
&= \sum_i \sum_{s_i \in S_i} \cdots \sum_{s_n \in S_n} \frac{dp_i^t}{dt}(s_i)\left(\prod_{j \neq i} p_j^t(s_j)\right)\Pi(s) \\
&= \sum_i \Pi\left(\frac{dp_i^t}{dt}, p_{-i}^t\right).
\end{aligned}
$$

The preceding explicit derivation essentially follows from the fact that Π is linear in its arguments, because these are mixed strategies of players. Therefore, the time derivative can be directly applied to the arguments. Now, observe that

$$\Pi\left(\frac{dp_i^t}{dt}, p_{-i}^t\right) = \Pi(\sigma_i^t - p_i^t, p_{-i}^t) = \Pi(\sigma_i^t, p_{-i}^t) - \Pi(p^t) = U_i(p^t),$$

where the second equality again follows by the linearity of Π in mixed strategies. The last equality uses the fact that $\sigma_i^t \in BR_i(p_{-i}^t)$. Combining this relation with the previous one, we have

$$\frac{d}{dt}(\Pi(p^t)) = \sum_i U_i(p^t) = W(t).$$

Since $W(t)$ is nonnegative everywhere, we conclude $\Pi(p^t)$ is nondecreasing as t increases; thus $\Pi^* = \lim_{t \to \infty} \Pi(p^t)$ exists (since Π is bounded above, $\Pi^* < \infty$). Moreover, we have

$$\Pi^* - \Pi(p^t) \geq \Pi(p^{t+\Delta}) - \Pi(p^t) = \int_0^\Delta W(t+\tau)d\tau \geq 0,$$

where the first inequality uses the fact that Π is nondecreasing; the middle inequality follows from the fundamental theorem of calculus, and the last inequality simply uses the fact that $W(t)$ is everywhere nonnegative. Since the left-hand side converges to zero, we conclude that $W(t) \to 0$ as $t \to \infty$. This establishes that for each i and for any initial condition p^0,

$$\lim_{t \to \infty}\left[\max_{\sigma_i' \in \Sigma_i} \Pi(\sigma_i', p_{-i}^t) - \Pi(p_i^t, p_{-i}^t)\right] = 0.$$

$\square$

2.3   GAMES WITH SPECIAL STRUCTURE

This section studies two special classes of games with appealing equilibrium and dynamic properties: supermodular games and potential games.

2.3.1   SUPERMODULAR GAMES

Supermodular games are games that are characterized by "strategic complementarities". Informally, this means that the marginal utility of increasing a player's strategy raises with increases in the other players' strategies. The implication is that the best response function of a player is a nondecreasing function of other players' strategies.

The machinery needed to study supermodular games is lattice theory and monotonicity results in lattice programming (see Appendix 2.A for more on lattices). The methods used are non-topological, and they exploit order properties. The existence of a pure Nash equilibrium in supermodular games can be established through a lattice-theoretic fixed point theorem, *Tarski's fixed point theorem*, which shows the existence of a fixed point for increasing functions (see [46] and Appendix 2.A for a statement of Tarski's fixed point theorem). Since for supermodular games the best-response correspondences of players have a monotone increasing selection, this implies existence of a pure Nash equilibrium. Below, we present an analysis that does not rely on Tarski's fixed point theorem, but instead uses ideas from the monotonicity results in lattice programming. This analysis also provides more insight into the structure of the equilibrium set in supermodular games.

Supermodular games are interesting for a number of reasons:

- They arise in many network models.

- We can establish the existence of a pure strategy equilibrium without requiring the quasi-concavity of the payoff functions.

- Many solution concepts yield the same predictions.

- The equilibrium set has a smallest and a largest element and there exists a simple algorithm to compute these.

- They have nice sensitivity (or comparative statics) properties and behave well under a variety of learning rules.

Much of the theory is due to Topkis [101], [100], Vives [103], [104], Milgrom and Roberts [69], and Milgrom and Shannon [70].

2.3.1.1   Monotonicity of Optimal Solutions

We first study the monotonicity properties of optimal solutions of parametric optimization problems. In particular, we consider

$$x(t) = \arg\max_{x \in X} f(x, t),$$

where $f : X \times T \to \mathbb{R}$, $X \subset \mathbb{R}$, and T is some partially ordered set (we will mostly focus on $T \subseteq \mathbb{R}^K$ with the usual *vector order*, i.e., for some $x, y \in T$, $x \geq y$ if and only if $x_i \geq y_i$ for all $i = 1, \ldots, k$; see Appendix 2.A for more general partially ordered sets). We are interested in conditions under which we can establish that $x(t)$ is a nondecreasing function of t. We next define the key property of *increasing differences*.[3]

Definition 2.25 Let $X \subseteq \mathbb{R}$ and T be some partially ordered set. A function $f : X \times T \to \mathbb{R}$ has *increasing differences* in (x, t) if for all $x' \geq x$ and $t' \geq t$, we have

$$f(x', t') - f(x, t') \geq f(x', t) - f(x, t).$$

The preceding definition implies that if f has increasing differences in (x, t), then the incremental gain to choosing a higher x (i.e., $x' \geq x$ rather than x) is greater when t is higher. That is $f(x', t) - f(x, t)$ is nondecreasing in t. One may verify that the property of increasing differences is symmetric: an equivalent statement is that if $t' \geq t$, then $f(x, t') - f(x, t)$ is nondecreasing in x.

Note that the function f need not be nicely behaved for the above definition to hold, nor do X and T need to be intervals. For instance, we could have $X = \{0, 1\}$ and just a few parameter values, e.g., $T = \{0, 1, 2\}$. If, however, f is nicely behaved, we can rewrite increasing differences in terms of partial derivatives.

Lemma 2.26 *Let $X \subseteq \mathbb{R}$ and $T \subseteq \mathbb{R}^k$ for some k, a partially ordered set with the usual vector order. Let $f : X \times T \to \mathbb{R}$ be a twice continuously differentiable function. Then, the following statements are equivalent:*

(a) The function f has increasing differences in (x, t).

(b) For all $t' \geq t$ and all $x \in X$, we have

$$\frac{\partial f(x, t')}{\partial x} \geq \frac{\partial f(x, t)}{\partial x}.$$

(c) For all $x \in X$, $t \in T$, and all $i = 1, \ldots, k$, we have

$$\frac{\partial^2 f(x, t)}{\partial x \partial t_i} \geq 0.$$

---

[3]The following analysis can be extended to the case where X is a lattice (see Topkis [101], and Milgrom and Roberts [69] for these extensions).

Before presenting our key result about monotonicity of optimal solutions, we introduce some examples, which satisfy increasing differences property.

Example 2.27   Network Effects

A set $\mathcal{I}$ of users can use one of two products X and Y (e.g., Blu-ray and HD DVD). We use $B_i(J, k)$ to denote the payoff to i when a subset J of users use k and $i \in J$. There exists a *positive externality* if

$$B_i(J, k) \leq B_i(J', k), \qquad \text{when } J \subset J',$$

i.e., player i is better off if more users use the same technology as him. This leads to a strategic form game with actions $S_i = \{X, Y\}$. Let us define the order $Y \succeq X$, which induces a lattice structure. Given $s \in S$, let $X(s) = \{i \in \mathcal{I} \mid s_i = X\}$, $Y(s) = \{i \in \mathcal{I} \mid s_i = Y\}$. We define the payoff functions as

$$u_i(s_i, s_{-i}) \doteq \begin{cases} B_i(X(s), X) & \text{if } s_i = X, \\ B_i(Y(s), Y) & \text{if } s_i = Y \end{cases}$$

It can be verified that the payoff functions satisfy increasing differences.

Example 2.28   Wireless Power Control

We consider the problem of power control in cellular CDMA wireless networks. In a CDMA system, all users access the channel using orthogonal codes at the same time, utilizing the entire available frequency spectrum (unlike time division multiple access (TDMA) or frequency division multiple access (FDMA) schemes). Direct sequence CDMA systems require stringent power control in the uplink channel to reduce the residual interference between users due to use of semi-orthogonal spreading codes. In the current IS-95 CDMA standard, the power control is done centrally by the base-station to mitigate the "near-far" problem by equalizing the received power of all mobile units regardless of their distance from the receiver.

With emerging multimedia applications, power control will be increasingly aimed at achieving different quality of service (QoS) requirements of applications. In the presence of heterogeneity in QoS requests, the resource (power) allocation problem becomes nonstandard and traditional network optimization techniques cannot be applied directly to come up with efficient distributed methods that would yield optimal allocations. To address this problem, recent literature has used game-theoretic models for resource control among heterogeneous users. This literature uses the "utility-maximization" framework of market economics, to provide different access privileges to users with different QoS requirements in a distributed manner (see [54], [57]). In this framework, each user (or equivalently application) is represented by a utility function that is a measure of his preferences over transmission rates. The consideration of game-theoretic methods for resource allocation in wireless networks is relatively more recent; see Chapter 4 for an overview of *wireless network games*.

It has been well-recognized that in the presence of interference, the strategic interactions between the users is in some scenarios that of strategic complementarities (see [93], [10], and [50]). Below, we present a simple game model for power allocation and show that the payoff functions satisfy increasing differences property. A different model for power allocation in wireless networks will be considered in the sequel in Chapter 4.

- Let $L = \{1, 2, ..., n\}$ denote the set of wireless nodes and

$$\mathcal{P} = \prod_{i \in L} [P_i^{\min}, P_i^{\max}] \subset \mathbb{R}^n$$

denote the set of power vectors $p = [p_1, \ldots, p_n]$ such that each node $i \in L$ transmits at a power level p_i.

- Received signal-to-interference ratio (SINR) for each node i, $\gamma_i : \mathcal{P} \to \mathbb{R}$ is given by

$$\gamma_i(p) = \frac{p_i h_i}{\sigma^2 + \sum_{j \neq i, 1 \leq j \leq n} p_j h_j},$$

where σ^2 is the noise variance (assuming an additive white Gaussian noise channel), and h_i is the channel gain from mobile i to the base station.

- Each user is endowed with a function $f_i(\gamma_i)$ as a function of its SINR γ_i.

- The payoff function of each user represents a tradeoff between the payoff obtained by the received SINR and the power expenditure, and it takes the form

$$u_i(p_i, p_{-i}) = f_i(\gamma_i) - cp_i.$$

Assume that each function f_i satisfies the following assumption regarding its *coefficient of relative risk aversion*:

$$\frac{-\gamma_i f_i''(\gamma_i)}{f_i'(\gamma_i)} \geq 1, \qquad \forall \, \gamma_i \geq 0.$$

We show that for all $i = 1 \ldots, n$, the function $u_i(p_i, p_{-i})$ has increasing differences in (p_i, p_{-i}). For this, we check the partial derivatives (see Lemma 2.26):

$$\frac{\partial u_i}{\partial p_i}(p_i, p_{-i}) = f'(\gamma_i)\frac{\gamma_i}{p_i} - c,$$

and for all $j \neq i$,

$$\frac{\partial^2 u_i}{\partial p_i \partial p_j}(p_i, p_{-i}) = -\frac{\gamma_i^2 h_j}{p_i^2 h_i}\left[\gamma_i f''(\gamma_i) + f'(\gamma_i)\right].$$

In view of our assumption on the coefficient of relative risk aversion of the utility functions f_i, i.e., $\frac{-\gamma_i f_i''(\gamma_i)}{f_i'(\gamma_i)} \geq 1$, it follows that $\frac{\partial^2 u_i}{\partial p_i \partial p_j}(p_i, p_{-i}) \geq 0$ for all $j \neq i$, showing the desired claim.

Consider the following class of utility functions:

$$f(\gamma) = \frac{\gamma^{1-\alpha}}{1 - \alpha}, \qquad \alpha > 1.$$

It can be seen that

$$\frac{-\gamma_i f_i''(\gamma_i)}{f_i'(\gamma_i)} = \alpha > 1.$$

Example 2.29   Oligopoly Models

We first consider Bertrand competition: Suppose firms $1, \ldots, I$ simultaneously choose prices, and the demand function is given by

$$D_i(p_i, p_{-i}) = a_i - b_i p_i + \sum_{j \neq i} d_{ij} p_j,$$

where b_i and d_{ij} are nonnegative constants. Let the strategy space be $S_i = [0, \infty)$ and the payoff function be $u_i(p_i, p_{-i}) = (p_i - c_i)D_i(p_i, p_{-i})$ (where as usual c_i is the cost of producing one unit of good). Then, $\frac{\partial^2 u_i}{\partial p_i \partial p_j}(p_i, p_{-i}) = d_{ij} \geq 0$, showing that $u_i(p_i, p_{-i})$ has increasing differences in (p_i, p_{-i}).

We next consider Cournot competition in a duopoly: two firms choose the quantity they produce $q_i \in [0, \infty)$. We denote the inverse demand function by $P(q_i, q_j)$, and assume that it is a function of $Q = q_i + q_j$, and it is twice continuously differentiable in Q. We further assume that

$$P'(Q) + q P''(Q) \leq 0.$$

Let the payoff function of each firm be $u_i(q_i, q_j) = q_i P(q_i + q_j) - cq_i$. Then, it can be seen that the payoff functions of the transformed game defined by $s_1 = q_1, s_2 = -q_2$ has increasing differences in (s_1, s_2).

2.3.1.2   Main Result

The next theorem presents the key result of our development, which is due to Topkis [101].

Theorem 2.30   *Let $X \subset \mathbb{R}$ be a compact set and T be some partially ordered set. Assume that the function $f : X \times T \to \mathbb{R}$ is upper semicontinuous in x for all $t \in T$ and has increasing differences in (x, t). Define $x(t) = \arg\max_{x \in X} f(x, t)$. Then, we have:*

1. *For all $t \in T$, $x(t)$ is nonempty and has a greatest and least element, denoted by $\bar{x}(t)$ and $\underline{x}(t)$, respectively.*

2. *For all $t' \geq t$, we have $\bar{x}(t') \geq \bar{x}(t)$ and $\underline{x}(t') \geq \underline{x}(t)$.*

Proof. (1) By the assumptions that for all $t \in T$, the function $f(\cdot, t)$ is upper semicontinuous and X is compact, it follows by the Weierstrass' Theorem that $x(t)$ is nonempty. For all $t \in T, x(t) \subset X$, therefore is bounded. Since $X \subset \mathbb{R}$, to establish that $x(t)$ has a greatest and lowest element, it suffices to show that $x(t)$ is closed.

Let $\{x^k\}$ be a sequence in $x(t)$. Since X is compact, x^k has a limit point $\bar{x}$. By restricting to a subsequence if necessary, we may assume without loss of generality that x^k converges to $\bar{x}$. Since $x^k \in x(t)$ for all k, we have

$$f(x^k, t) \geq f(x, t), \qquad \forall x \in X.$$

Taking the limit as $k \to \infty$ in the preceding relation and using the upper semicontinuity of $f(\cdot, t)$, we obtain

$$f(\bar{x}, t) \geq \limsup_{k \to \infty} f(x^k, t) \geq f(x, t), \qquad \forall x \in X,$$

thus showing that $\bar{x}$ belongs to $x(t)$, and proving the desired closedness claim.

(2) Let $t' \geq t$. Let $x \in x(t)$ and $x' = \bar{x}(t')$. By the fact that x maximizes $f(x, t)$, we have

$$f(x, t) - f(\min(x, x'), t) \geq 0.$$

This implies (by verifying the two cases: $x \geq x'$ and $x' \geq x$) that

$$f(\max(x, x'), t) - f(x', t) \geq 0.$$

By increasing differences of f, this yields

$$f(\max(x, x'), t') - f(x', t') \geq 0.$$

Thus, $\max(x, x')$ maximizes $f(\cdot, t')$, i.e, $\max(x, x')$ belongs to $x(t')$. Since x' is the greatest element of the set $x(t')$, we conclude that $\max(x, x') \leq x'$, thus $x \leq x'$. Since x is an arbitrary element of $x(t)$, this implies $\bar{x}(t) \leq \bar{x}(t')$. A similar argument applies to the lowest maximizers.

$\square$

The above theorem suggests that if f has increasing differences, then the set of maximizers $x(t)$ is nondecreasing in t, in the sense that both the greatest maximizers $\bar{x}(t)$ and the lowest maximizers $\underline{x}(t)$ are nondecreasing in t.

2.3.1.3 Supermodular Games

We now introduce the class of supermodular games.

Definition 2.31 Supermodular game   The strategic form game $\langle \mathcal{I}, (S_i), (u_i) \rangle$ is a *supermodular game* if for all i:

1. S_i is a compact subset of $\mathbb{R}$ (or more generally S_i is a sublattice of $\mathbb{R}^m$),

2. u_i is upper semicontinuous in s_i, continuous in s_{-i},

3. u_i has increasing differences in (s_i, s_{-i}).[4]

Applying Topkis' theorem in this context immediately implies that each player's best response correspondence is increasing in the actions of other players.

Corollary 2.32   *Assume $\langle \mathcal{I}, (S_i), (u_i) \rangle$ is a supermodular game. Let*

$$B_i(s_{-i}) = \arg \max_{s_i \in S_i} u_i(s_i, s_{-i}).$$

Then:

1. *$B_i(s_{-i})$ has a greatest and least element, denoted by $\bar{B}_i(s_{-i})$ and $\underline{B}_i(s_{-i})$.*

2. *If $s'_{-i} \geq s_{-i}$, then $\bar{B}_i(s'_{-i}) \geq \bar{B}_i(s_{-i})$ and $\underline{B}_i(s'_{-i}) \geq \underline{B}_i(s_{-i})$.*

We now use the properties of the supermodular games to show that various solution concepts that we considered in Chapter 1 for strategic form games yield the same predictions in supermodular games.

Theorem 2.33   *Let $\langle \mathcal{I}, (S_i), (u_i) \rangle$ be a supermodular game. Then the set of strategies that survive iterated strict dominance in pure strategies (i.e., iterated elimination of strictly dominated pure strategies, see Section 1.2.2) has greatest and least elements $\bar{s}$ and $\underline{s}$, coinciding with the greatest and the least pure strategy Nash Equilibria.*

The preceding theorem immediately yields the following corollary.

Corollary 2.34   *Supermodular games have the following properties:*

1. *Pure strategy NE exist.*

2. *The largest and smallest strategies that are compatible with iterated strict dominance (ISD), correlated equilibrium, and Nash equilibrium are the same.*

3. *If a supermodular game has a unique NE, it is dominance solvable.*[5]

We now return to the proof of Theorem 2.33.

---

[4]More generally, u_i is supermodular in (s_i, s_{-i}); see Fudenberg and Tirole [46] for the more general definition of *supermodularity*, which is an extension of the property of increasing differences to games with multi-dimensional strategy spaces.

[5]Consequently, several learning and adjustment rules, such as best-response dynamics, converge to it.

Proof. We iterate the best response mapping. Let $S^0 = S$, and let $s^0 = (s_1^0, \ldots, s_I^0)$ be the largest element of S. Let $s_i^1 = \bar{B}_i(s_{-i}^0)$ and $S_i^1 = \{s_i \in S_i^0 \mid s_i \leq s_i^1\}$. We show that any $s_i > s_i^1$, i.e, any $s_i \notin S_i^1$, is strictly dominated by s_i^1. For all $s_{-i} \in S_{-i}$, we have

$$
\begin{aligned}
u_i(s_i, s_{-i}) - u_i(s_i^1, s_{-i}) &\leq u_i(s_i, s_{-i}^0) - u_i(s_i^1, s_{-i}^0) \\
&< 0,
\end{aligned}
$$

where the first inequality follows by the increasing differences of $u_i(s_i, s_{-i})$ in (s_i, s_{-i}), and the strict inequality follows by the fact that s_i is not a best response to s_{-i}^0. Note that $s_i^1 \leq s_i^0$. Iterating this argument, We define

$$
s_i^k = \bar{B}_i(s_{-i}^{k-1}), \qquad S_i^k = \{s_i \in S_i^{k-1} \mid s_i \leq s_i^k\}.
$$

Assume $s^k \leq s^{k-1}$. Then, by Corollary 2.32, we have

$$
s_i^{k+1} = \bar{B}_i(s_{-i}^k) \leq \bar{B}_i(s_{-i}^{k-1}) = s_i^k.
$$

This shows that the sequence $\{s_i^k\}$ is a decreasing sequence, which is bounded from below, and hence it has a limit, which we denote by $\bar{s}_i$. Only the strategies $s_i \leq \bar{s}_i$ are undominated.

Similarly, we can start with $s^0 = (s_1^0, \ldots, s_I^0)$ the smallest element in S and identify $\underline{s}$.

To complete the proof, we show that $\bar{s}$ and $\underline{s}$ are NE. By construction, for all i and $s_i \in S_i$, we have

$$
u_i(s_i^{k+1}, s_{-i}^k) \geq u_i(s_i, s_{-i}^k).
$$

Taking the limit as $k \to \infty$ in the preceding relation and using the upper semicontinuity of u_i in s_i and continuity of u_i in s_{-i}, we obtain

$$
u_i(\bar{s}_i, \bar{s}_{-i}) \geq u_i(s_i, \bar{s}_{-i}),
$$

showing the desired claim. $\qquad\square$

2.3.2   POTENTIAL GAMES

In this section, we present the family of potential games, introduced by Monderer and Shapley [76], and study their properties. A strategic form game is a *potential game* if there exists a function $\Phi : S \to \mathbb{R}$ such that $\Phi(s_i, s_{-i})$ gives information about the utility functions $u_i(s_i, s_{-i})$ for each $i \in \mathcal{I}$. The function Φ is referred to as the *potential function*. The potential function has a natural analogy to "energy" in physical systems. It will be useful both for locating pure strategy Nash equilibria and also for the analysis of "myopic" dynamics.

We next formally define different classes of potential games. Let $G = \langle \mathcal{I}, (S_i), (u_i) \rangle$ be a strategic form game.

Definition 2.35   A function $\Phi : S \to \mathbb{R}$ is called an *ordinal potential function* for the game G if for each $i \in \mathcal{I}$ and all $s_{-i} \in S_{-i}$,

$$
u_i(x, s_{-i}) - u_i(z, s_{-i}) > 0 \text{ iff } \Phi(x, s_{-i}) - \Phi(z, s_{-i}) > 0, \text{ for all } x, z \in S_i.
$$

G is called an *ordinal potential game* if it admits an ordinal potential.

Definition 2.36   A function $\Phi : S \to \mathbb{R}$ is called an *(exact) potential function* for the game G if for each $i \in \mathcal{I}$ and all $s_{-i} \in S_{-i}$,

$$u_i(x, s_{-i}) - u_i(z, s_{-i}) = \Phi(x, s_{-i}) - \Phi(z, s_{-i}), \text{ for all } x, z \in S_i.$$

G is called an *(exact) potential game* if it admits a potential.

A potential function assigns a real value for every $s \in S$. Thus, for bimatrix games (i.e., finite games with two players), we can also represent the potential function as a matrix, each entry corresponding to the vector of strategies from the payoff matrix.

Example 2.37   The matrix P represents a potential function for the "Prisoner's dilemma" game described below:

$$G = \begin{pmatrix} (1,1) & (9,0) \\ (0,9) & (6,6) \end{pmatrix}, \qquad P = \begin{pmatrix} 4 & 3 \\ 3 & 0 \end{pmatrix}$$

The next theorem establishes the existence of a pure strategy Nash equilibrium in finite ordinal potential games.

Theorem 2.38   *Every finite ordinal potential game has at least one pure strategy Nash equilibrium.*

Proof. The global maximum of an ordinal potential function is a pure strategy Nash equilibrium. To see this, suppose that s^* corresponds to the global maximum. Then, for any $i \in \mathcal{I}$, we have, by definition, $\Phi(s_i^*, s_{-i}^*) - \Phi(s, s_{-i}^*) \geq 0$ for all $s \in S_i$. But since Φ is a potential function, for all i and all $s \in S_i$,

$$u_i(s_i^*, s_{-i}^*) - u_i(s, s_{-i}^*) \geq 0 \quad \text{iff} \quad \Phi(s_i^*, s_{-i}^*) - \Phi(s, s_{-i}^*) \geq 0.$$

Therefore, $u_i(s_i^*, s_{-i}^*) - u_i(s, s_{-i}^*) \geq 0$ for all $s \in S_i$ and for all $i \in \mathcal{I}$. Hence, s^* is a pure strategy Nash equilibrium. $\qquad\square$

Note, however, that there may also be other pure strategy Nash equilibria corresponding to local maxima of the potential function. The next two examples present examples of ordinal and exact potential games.

Example 2.39 Cournot Competition I firms choose quantity $q_i \in [0, \infty)$. The payoff function for player i given by $u_i(q_i, q_{-i}) = q_i(P(Q) - c)$. We define the function

$$\Phi(q_1, \cdots, q_I) = \left(\prod_{i=1}^{I} q_i \right)(P(Q) - c).$$

Note that for all i and all $q_{-i} > 0$,

$$u_i(q_i, q_{-i}) - u_i(q_i', q_{-i}) > 0 \text{ iff } \Phi(q_i, q_{-i}) - \Phi(q_i', q_{-i}) > 0, \ \forall \, q_i, q_i' > 0.$$

The function Φ is therefore an ordinal potential function for this game.

Example 2.40   Cournot Competition   Suppose now that $P(Q) = a - bQ$ and costs $c_i(q_i)$ are arbitrary. We define the function

$$\Phi^*(q_1, \cdots, q_n) = a \sum_{i=1}^{I} q_i - b \sum_{i=1}^{I} q_i^2 - b \sum_{1 \leq i < l \leq I} q_i q_l - \sum_{i=1}^{I} c_i(q_i).$$

It can be shown that for all i and all q_{-i},

$$u_i(q_i, q_{-i}) - u_i(q_i', q_{-i}) = \Phi^*(q_i, q_{-i}) - \Phi^*(q_i', q_{-i}), \text{ for all } q_i, q_i' > 0.$$

The function Φ is an exact potential function for this game.

We next study the convergence behavior of simple myopic dynamics in finite ordinal potential games. The next definition formalizes the strategy profile trajectories generated by such dynamics in potential games.

Definition 2.41   A *path* in strategy space S is a sequence of strategy vectors $(s^0, s^1, \cdots)$ such that every two consecutive strategies differ in one coordinate (i.e., exactly in one player's strategy). An *improvement path* is a path $(s^0, s^1, \cdots)$ such that, $u_{i_k}(s^k) < u_{i_k}(s^{k+1})$ where s^k and s^{k+1} differ in the i_k^{th} coordinate. In other words, the payoff improves for the player who changes his strategy.

An improvement path can be thought of as generated dynamically by "myopic players", who update their strategies according to *one-sided better reply dynamic*.

Proposition 2.42   *In every finite ordinal potential game, every improvement path is finite.*

Proof.   Suppose $(s^0, s^1, \cdots)$ is an improvement path. Therefore, we have,

$$\Phi(s^0) < \Phi(s^1) < \cdots,$$

where Φ is the ordinal potential. Since the game is finite, i.e., it has a finite strategy space, the potential function takes on finitely many values and the above sequence must end in finitely many steps.   □

This result implies that in finite ordinal potential games, every "maximal" improvement path must terminate in an equilibrium point. That is, the simple myopic learning process based on one-sided better reply dynamic converges to the equilibrium set.

We conclude this section by presenting a widely applicable class of games, congestion games, introduced by Rosenthal [89]. Congestion games are characterized by resources shared among many agents, which experience negative externalities due to the resulting congestion effects. Hence, this class of games is a standard model of strategic interactions in many network games. We show that congestion games are potential games and therefore share the appealing properties of these games.

2.3.2.1 Congestion Games

We define the congestion model as a tuple $C = \langle \mathcal{I}, \mathcal{M}, (S_i)_{i \in \mathcal{I}}, (c^j)_{j \in \mathcal{M}} \rangle$ where:

- $\mathcal{I} = \{1, 2, \cdots, I\}$ is the set of players.

- $\mathcal{M} = \{1, 2, \cdots, m\}$ is the set of resources.

- S_i is the set of nonempty resource combinations (e.g., links or common resources) that player i can take or use. A strategy for player i is $s_i \in S_i$, corresponding to the subset of resources that this player is using.

- $c^j(k)$ is the benefit for the negative of the cost to each user who uses resource j if k users are using it.

Given this model, we define the congestion game $\langle \mathcal{I}, (S_i), (u_i) \rangle$ with utilities

$$u_i(s_i, s_{-i}) = \sum_{j \in s_i} c^j(k_j),$$

where k_j is the number of users of resource j under strategy s.

The next theorem shows that every congestion game is a potential game.

Theorem 2.43 *Every congestion game is a potential game and thus has a pure strategy Nash equilibrium.*

Proof. For each j define $\bar{k}_j^i$ as the usage of resource j excluding player i, i.e.,

$$\bar{k}_j^i = \sum_{i' \neq i} \mathbf{I}[j \in s_{i'}],$$

where $\mathbf{I}[j \in s_{i'}]$ is the indicator for the event that $j \in s_{i'}$. With this notation, the utility difference of player i from two strategies s_i and s_i' (when others are using the strategy profile s_{-i}) is

$$u_i(s_i, s_{-i}) - u_i(s_i', s_{-i}) = \sum_{j \in s_i} c^j(\bar{k}_j^i + 1) - \sum_{j \in s_i'} c^j(\bar{k}_j^i + 1).$$

Now consider the function

$$\Phi(s) = \sum_{j \in \bigcup_{i' \in \mathcal{I}} s_{i'}} \left[\sum_{k=1}^{k_j} c^j(k) \right].$$

We can also write

$$\Phi(s_i, s_{-i}) = \sum_{\substack{j \in \bigcup s_{i'} \\ i' \neq i}} \left[\sum_{k=1}^{\bar{k}^i_j} c^j(k) \right] + \sum_{j \in s_i} c^j(\bar{k}^i_j + 1).$$

Therefore, we have

$$\Phi(s_i, s_{-i}) - \Phi(s'_i, s_{-i}) = \sum_{\substack{j \in \bigcup s_{i'} \\ i' \neq i}} \left[\sum_{k=1}^{\bar{k}^i_j} c^j(k) \right] + \sum_{j \in s_i} c^j(\bar{k}^i_j + 1)$$

$$- \sum_{\substack{j \in \bigcup s_{i'} \\ i' \neq i}} \left[\sum_{k=1}^{\bar{k}^i_j} c^j(k) \right] + \sum_{j \in s'_i} c^j(\bar{k}^i_j + 1)$$

$$= \sum_{j \in s_i} c^j(\bar{k}^i_j + 1) - \sum_{j \in s'_i} c^j(\bar{k}^i_j + 1)$$

$$= u_i(s_i, s_{-i}) - u_i(s'_i, s_{-i}).$$

$\square$

2.A   APPENDIX: LATTICES

Let $\geq$ be a binary relation on a nonempty set S. The pair $(S, \geq)$ is a *partially ordered set* if $\geq$ is reflexive ($x \geq x$ for all $x \in S$), transitive ($x \geq y$ and $y \geq z$ implies that $x \geq z$), and antisymmetric ($x \geq y$ and $y \geq x$ implies that $x = y$). A partially ordered set $(S, \geq)$ is *(completely) ordered* if for $x \in S$ and $y \in S$, either $x \geq y$ or $y \geq x$.

A *lattice* is a partially ordered set $(S, \geq)$ in which any two elements x, y have a least upper bound (supremum), $\sup_S(x, y) = \inf\{z \in S \mid z \geq x, z \geq y\}$[6], and a greatest lower bound (infimum), $\inf_S(x, y) = \sup\{z \in S \mid z \leq x, z \leq y\}$[7], in the set. For example, any interval of the real line with the usual order is a lattice since any two points have a supremum and infimum in the interval. However, the set $S \subset \mathbb{R}^2$, $S = \{(1, 0), (0, 1)\}$ is not a lattice with the vector ordering (the usual component-wise ordering: $x \leq y$ if and only if $x_i \leq y_i$ for any i) since $(1, 0)$ and $(0, 1)$ have no joint upper bound in S. The set $S' = \{(0, 0), (0, 1), (1, 0), (1, 1)\}$ is indeed a lattice with the vector ordering.

---

[6]Supremum of $\{x, y\}$ is denoted by $x \vee y$ and is called the *join* of x and y.
[7]Infimum of $\{x, y\}$ is denoted by $x \wedge y$ and is called the *meet* of x and y.

Similarly, the simplex in $\mathbb{R}^n$ (again with the usual vector ordering) $\{x \in \mathbb{R}^n \mid \sum_i x_i = 1, \ x_i \geq 0\}$ is not a lattice, while the box $\{x \in \mathbb{R}^n \mid 0 \leq x_1 \leq 1\}$ is.

A lattice $(S, \geq)$ is *complete* if every nonempty subset of S has a supremum and an infimum in S. Any compact interval of the real line with the usual order is a complete lattice, while the open interval (a, b) is a lattice but is not complete (indeed, the supremum of (a, b) does not belong to (a, b)).

A subset L of the lattice S is a *sublattice* of S if the supremum and infimum of any two elements of L (with the supremum and infimum is taken with respect to S) belong also to L. That is, a sublattice L of the lattice S is a subset of S that is closed under the operations of supremum and infimum. The sublattice L of S is *complete* if every nonempty subset of L has a supremum and infimum in L. A subset that is a lattice or complete lattice in its own right may not be a sublattice or complete sublattice of a larger lattice, because the relevant suprema and infima are defined relative to the larger lattice. Thus, the set $T = [0, 1) \cup \{2\}$ is a complete lattice under the usual ordering; the supremum in T for the set $[0, 1)$ is $2 \in T$. However, T is not a complete sublattice of the lattice $[0,2]$, because then $\sup[0, 1) = 1 \notin T$.

A function $f : S \to \mathbb{R}$ is *supermodular on S* if for all $x, y \in S$

$$f(x) + f(y) \leq f(x \wedge y) + f(x \vee y).$$

Note that supermodularity is automatically satisfied if S_i is single dimensional. We next provide an alternative characterization of supermodularity for smooth functions.

Theorem 2.44 *Let I be an interval $\mathbb{R}^n$. Assume that $f : \mathbb{R}^n \to \mathbb{R}$ is twice continuously differentiable on some open set containing I. Then f is supermodular on I if and only if for all $x \in I$ and all $i \neq j$,*

$$\frac{\partial^2 f}{\partial x_i \partial x_j} \geq 0.$$

In general, supermodularity uses only the order structure of the lattice. It entails no assumptions of convexity or connectedness of the domain, nor does it require any convexity, differentiability of the function itself. However, in view of this theorem it is easy to check whether smooth functions on Euclidean intervals are supermodular.

Let $(S, \geq)$ be a partially ordered set. A function f from S to S is *increasing* if for all $x, y \in S$, $x \geq y$ implies $f(x) \geq f(y)$.

We finally state the following lattice-theoretical fixed point theorem due to Tarski.

Theorem 2.45 Tarski *Let $(S, \geq)$ be a complete lattice and $f : S \to S$ an increasing function. Then, the set of fixed points of f, denoted by E, is nonempty and $(E, \geq)$ is a complete lattice.*

PART II

Network Games

CHAPTER 3

Wireline Network Games

Many games are played over *networks*, in the sense that players' payoffs depend on others through a network-like structure. Classical examples are the allocation of network flows in a communication network, or of traffic in a transportation network. Since the routing decisions of each user usually affect the performance of other users (through commonly shared links), this leads to a *noncooperative routing game* among the users. Our focus in this chapter is on the equilibria of such routing games. In particular, we shall consider the important engineering aspect of the *efficiency* of the equilibrium, namely how "good" the equilibrium point is compared to the socially optimal operating point (with respect to a properly defined quality measure).

In Section 3.1, we formally define a general routing model and the associated equilibrium concepts. We demonstrate that inefficiencies might occur at equilibrium due to selfish behavior. Moreover, due to self-interested decisions of users, we show that the addition of network resources may deteriorate network performance, a phenomenon known as the *Braess' paradox* [27].

The basic routing game ignores the service provider role in routing traffic. Most large-scale communication networks, such as the Internet, consist of interconnected administrative domains. While source (or selfish) routing, where transmission follows the least cost path for each source, is reasonable across domains, service providers typically engage in traffic engineering to improve operating performance within their own network. Motivated by this observation, we develop and analyze in Section 3.2 a model of *partially optimal routing*, where optimal routing within subnetworks is overlaid with selfish routing across domains.

Another aspect that is missing from a basic routing game is the possibility that a service provider charges a price for utilizing its resources. Prices are often set by multiple service providers in control of their administrative domains with the objective of maximizing their (long-run) revenues. In Section 3.3, we investigate the implications of profit-maximizing pricing by multiple decentralized service providers. Finally, several extensions to the models considered in this chapter are outlined in Section 3.4.

3.1 SELFISH ROUTING, WARDROP EQUILIBRIUM AND EFFICIENCY

Before precisely formulating the routing model that will be the subject of this chapter, consider the motivating network example depicted in Figure 3.1. The example in this figure is due to Pigou (1920). There is one unit of load that needs to be routed from a source node to a destination node. This load corresponds to aggregate load of infinitesimal users (e.g., motorists in a transportation network).

There are two alternative links (routes) that may carry the traffic. Each link is characterized by a per-unit cost function. In the transportation context, the cost may correspond to the (average) delay of a vehicle[1]. Note that the cost in the upper link depends on the amount of load it accommodates, while the load in the lower link is a constant, thus congestion independent.

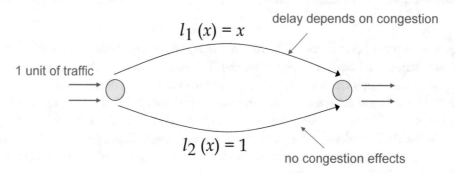

Figure 3.1: The Pigou example.

Observe first that the socially optimal solution, i.e., the flow allocation which minimizes the aggregate cost is to split traffic equally between the two routes, giving

$$\min_{x_1+x_2\leq 1} C_{\text{system}}(x^S) = \sum_i l_i(x_i^S)x_i^S = \frac{1}{4} + \frac{1}{2} = \frac{3}{4}.$$

However, in a noncooperative framework, the above solution does not correspond to a Nash equilibrium. Indeed, consider the corresponding game between the infinitesimal players (such games between an infinite number of "small" users are often referred to as *non-atomic games*). The players that use the bottom link experience a cost which is strictly higher than the cost experienced by players in the upper link, and would therefore modify their routing decision. Consequently, the unique Nash equilibrium of the game, also referred to as *Wardrop equilibrium* (WE)[2] is $x_1 = 1$ and $x_2 = 0$ (since for any $x_1 < 1, l_1(x_1) < 1 = l_2(1 - x_1)$), giving an aggregate cost of

$$C_{\text{eq}}(x^{WE}) = \sum_i l_i(x_i^{WE})x_i^{WE} = 1 + 0 = 1.$$

Note that the aggregate cost at the Wardrop equilibrium is larger than the optimal cost by a factor of $\frac{4}{3}$. We will establish later in Section 3.1.3 that $\frac{4}{3}$ is actually the worst-case factor for the efficiency loss when all network topologies are taken into account (as long as the delay costs are linear, as in our example).

The rest of this section is organized as follows. We present the routing model in Section 3.1.1, restricting attention to a single source-destination pair. We define and characterize the Wardrop

---

[1]Accordingly, the terms "delay" and "cost" are interchangeably used in this chapter.
[2]Named after John Glen Wardrop, an English transport analyst who developed this equilibrium concept in 1952 [105].

equilibrium in Section 3.1.2. We then study in Section 3.1.3 the efficiency loss incurred by self-interested behavior. We conclude this section by extending the framework of the routing problem to multiple source-destinations.

3.1.1   ROUTING MODEL

We consider in this section networks with a single origin-destination pair. Let $N = (V, A)$ be the directed network with V being the set of nodes and A being the set of links. Denote by $\mathcal{P}$ the set of paths between the origin and the destination. Let x_p denote the flow on path $p \in \mathcal{P}$. We assume that each link $i \in A$ has a latency function $l_i(x_i)$, where

$$x_i = \sum_{\{p \in \mathcal{P} | i \in p\}} x_p. \tag{3.1}$$

Here the notation $p \in \mathcal{P} | i \in p$ denotes the paths p that traverse link $i \in A$. The latency function captures congestion effects, hence the latency is a function of the *total* flow on the link. We assume for simplicity that $l_i(x_i)$ is nonnegative, differentiable, and nondecreasing. We further assume that all traffic is homogeneous, in the sense that all players (e.g., drivers) have the same experience when utilizing the link. The total traffic is normalized to one, and the set of players is accordingly given by $\mathcal{I} = [0, 1]$.

Remark 3.1   So far in this monograph, we took the set of players, $\mathcal{I}$, to be a finite set. Nonetheless, this is not essential for a proper definition of a game. In non-atomic games, $\mathcal{I}$ is typically taken to be some interval in $\mathbb{R}$, as in the above.

We denote a routing pattern by the vector $\mathbf{x}$. If $\mathbf{x}$ satisfies (3.1), and furthermore $\sum_{p \in \mathcal{P}} x_p = 1$ and $x_p \geq 0$ for all $p \in \mathcal{P}$, then $\mathbf{x}$ is a feasible routing pattern. The total delay (latency) cost of a routing pattern $\mathbf{x}$ is:

$$C(\mathbf{x}) = \sum_{i \in A} x_i l_i(x_i).$$

That is, it is the sum of latencies $l_i(x_i)$ for each link $i \in A$ multiplied by the flow over this link, x_i, summed over all links A.

The *socially optimal routing* x^S is a feasible routing pattern that minimizes the aggregate cost; it can be obtained as a solution of the following optimization problem

$$
\begin{aligned}
&\text{minimize} && \sum_{i \in A} x_i l_i(x_i) \\
&\text{subject to} && \sum_{\{p \in \mathcal{P} | i \in p\}} x_p = x_i, \text{ for all } i \in E, \\
& && \sum_{p \in \mathcal{P}} x_p = 1 \text{ and } x_p \geq 0 \text{ for all } p \in \mathcal{P}.
\end{aligned}
$$

3.1.2   WARDROP EQUILIBRIUM

As indicated above, an underlying assumption in our routing model is that each player is "infinitesimal", i.e., has a negligible effect on overall performance. A Wardrop equilibrium, which we formally define below, is a convenient modeling tool when each participant in the game is small. This equilibrium notion can be regarded as a Nash equilibrium in this game, where the strategies of the other players are replaced by *aggregates*, due to the non-atomic nature of the game; in our specific context, the aggregates correspond to the total traffic on different routes.

We now proceed to formally define the notion of a Wardrop equilibrium. For concreteness, we define the equilibrium in the context of a transportation network, although the context can obviously be more general, and correspond to any non-atomic game.

Since a Wardrop equilibrium can be viewed as a Nash equilibrium with an infinite number of small decision makers, it has to be the case that for each motorist their routing choice must be optimal. This implies that if a motorist $k \in \mathcal{I}$ is using path p, then there does not exist path p' such that

$$\sum_{i \in p} l_i(x_i) > \sum_{i \in p'} l_i(x_i).$$

A Wardrop equilibrium is formally defined as follows.

Definition 3.2   A feasible flow patters $\mathbf{x}$ is a *Wardrop equilibrium* if

$$\sum_{i \in p'} l_i(x_i) = \sum_{i \in p} l_i(x_i) \text{ for all } p, p' \in \mathcal{P} \text{ with } x_p, x_{p'} > 0, \text{ and}$$

$$\sum_{i \in p'} l_i(x_i) \geq \sum_{i \in p} l_i(x_i) \text{ For all } p, p' \in \mathcal{P} \text{ with } x_p > 0 \text{ and } x_{p'} = 0.$$

A fundamental property of the Wardrop equilibrium is that it can be obtained via the solution of a *convex* optimization problem. This important property suggests that a Wardrop equilibrium can be characterized (i.e., computed) efficiently. We state this result below and provide an outline for its proof.

Theorem 3.3   Beckmann, McGuire and Winsten [18]   *A feasible routing pattern* $\mathbf{x}^{WE}$ *is a Wardrop equilibrium if and only if it is a solution to*

$$\begin{aligned}
minimize \quad & \sum_{i \in A} \int_0^{x_i} l_i(z)\, dz \\
subject\ to \quad & \sum_{\{p \in \mathcal{P} | i \in p\}} x_p = x_i, \text{ for all } i \in A, \\
& \sum_{p \in \mathcal{P}} x_p = 1 \text{ and } x_p \geq 0 \text{ for all } p \in \mathcal{P}.
\end{aligned}$$
(3.2)

Moreover, if each l_i is strictly increasing, then $\mathbf{x}^{WE}$ is unique.

Proof. Observe first that by Weierstrass' Theorem, a solution to (3.2) exists, and thus a Wardrop equilibrium always exists. Next, note that (3.2) can be written as

$$\text{minimize} \quad \sum_{i \in A} \int_0^{\sum_{i \in p} x_p} l_i(z) \, dz$$

$$\text{subject to} \quad \sum_{p \in \mathcal{P}} x_p = 1 \text{ and } x_p \geq 0 \text{ for all } p \in \mathcal{P}.$$

Since each l_i is nondecreasing, this is a convex program. Therefore, first-order conditions are necessary and sufficient. The first-order conditions with respect to x_p are

$$\sum_{i \in p} l_i \left(x_i^{WE} \right) \geq \lambda$$

with the complementary slackness, i.e., with equality whenever $x_p^{WE} > 0$. Here λ is the Lagrange multiplier on the constraint $\sum_{p \in \mathcal{P}} x_p = 1$. Consequently, the Lagrange multiplier will be equal to the lowest cost path, which then implies the result that for all $p, p' \in \mathcal{P}$ with $x_p^{WE}, x_{p'}^{WE} > 0$, $\sum_{i \in p'} l_i(x_i^{WE}) = \sum_{i \in p} l_i(x_i^{WE})$; clearly, for paths with $x_p^{WE} = 0$, the cost can be higher.

Finally, if each l_i is strictly increasing, then the set of equalities $\sum_{i \in p'} l_i(x_i^{WE}) = \sum_{i \in p} l_i(x_i^{WE})$ admits unique solution, establishing uniqueness. □

3.1.3 INEFFICIENCY OF THE EQUILIBRIUM

We saw from the Pigou example that the Wardrop equilibrium fails to minimize total delay, hence it is generally inefficient when compared to the performance at the social optimum. More generally, it is well known that equilibria exhibit inefficiencies in diverse noncooperative scenarios. Koutsoupias and Papadimitriou [55] introduced the term *Price of Anarchy* (POA) to quantify these inefficiencies in games over all possible instances.

In our routing context, let $\mathcal{R}'$ denote the set of all routing instances, covering all possible network topologies and all latency functions which belong to a given family of functions (e.g., affine). Then the PoA is defined as the worst-case efficiency over all instances, namely

$$\inf_{R \in \mathcal{R}'} \frac{C(x^S(R))}{C(x^{WE}(R))}.$$

The Pigou example establishes that when restricting ourselves to affine latency functions, the PoA is at least $\frac{3}{4}$. The following theorem establishes that the PoA is exactly $\frac{3}{4}$, namely $\frac{3}{4}$ is the worst possible ratio between the social optimum cost and the cost at a Wardrop equilibrium.

Theorem 3.4   Roughgarden and Tardos [91]

   Let $\mathcal{R}^{conv}$ and $\mathcal{R}^{aff}$ denote the class of all routing instances where latency functions are convex and affine, respectively.

   (a)

$$\inf_{R \in \mathcal{R}^{conv}} \frac{C(x^{SO}(R))}{C(x^{WE}(R))} = 0.$$

   (b) Consider a routing instance $R = (V, A, P, s, t, X, 1)$ where l_j is an affine latency function for all $j \in A$. Then,

$$\frac{C(x^{SO}(R))}{C(x^{WE}(R))} \geq \frac{3}{4}.$$

   Furthermore, the bound above is tight.

Proof. We shall prove part (b) of the theorem later in Section 3.2.3 under a more general setup (Theorem 3.9).

   We prove part (a) by means of a simple example, demonstrating that the Wardrop equilibrium can be arbitrarily inefficient when allowing the larger set of convex increasing latency functions. Consider the example in Figure 3.2, which is the same as the Pigou example, except with a different latency on link 1 which is now $l_1(x) = x^k, k \geq 1$.

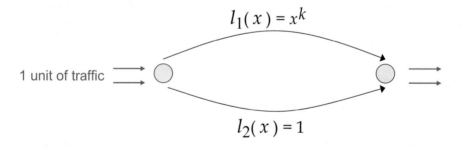

Figure 3.2: Extension of the Pigou example to non-linear latency functions.

   In this example, the socially optimal routing is obtained by the following equation

$$l_1(x_1) + x_1 l_1'(x_1) = l_2(1 - x_1) + (1 - x_1) l_2'(1 - x_1) ,$$

leading to

$$x_1^k + k x_1^k = 1 .$$

Therefore, the system optimum sets $x_1 = (1 + k)^{-1/k}$ and $x_2 = 1 - (1 + k)^{-1/k}$, so that

$$\min_{x_1 + x_2 \leq 1} C_{\text{system}}(x^S) = \sum_i l_i(x_i^S) x_i^S = (1 + k)^{-\frac{k+1}{k}} + 1 - (1 + k)^{-1/k} .$$

The Wardrop equilibrium again has $x_1 = 1$ and $x_2 = 0$ (since once again for any $x_1 < 1$, $l_1(x_1) < 1 = l_2(1 - x_1)$). Thus

$$C_{eq}(x^{WE}) = \sum_i l_i(x_i^{WE})x_i^{WE} = 1 + 0 = 1.$$

Therefore, the PoA is upper bounded by

$$\frac{C_{system}(x^S)}{C_{eq}(x^{WE})} = (1+k)^{-\frac{k+1}{k}} + 1 - (1+k)^{-1/k}.$$

This limits to 0 as $k \to \infty$ (the first term tends to zero, while the last term limits to one). Thus, the equilibrium can be *arbitrarily* inefficient relative to the social optimum.

There has been extensive research on the Price of Anarchy in related selfish routing models. We do not attempt to cover all models in this survey. We refer the reader to a survey book by Roughgarden [90] on the subject.

The Price of Anarchy notion allows one to quantify the extent to which selfish behavior affects network performance. A related consequence of selfish behavior in networks is captured through the celebrated Braess' Paradox [27], which demonstrates that the addition of an intuitively helpful route negatively impacts network users at equilibrium. Figure 3.3 depicts the example used by Braess. A link with zero cost is added to the network, nonetheless the equilibrium, which coincided with the social optimum before the addition of the new link, becomes inefficient; this can be immediately seen, as the overall cost increases. This outcome is paradoxical since the addition of another route should help traffic; obviously, the addition of a link can never increase aggregate delay in the social optimum. An active research direction has been to identify conditions under which the Braess' paradox would never occur, e.g., based on the network topology or other factors; see [90] and references therein for more details.

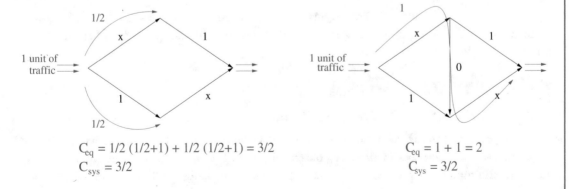

Figure 3.3: The Braess' Paradox.

3.1.4 MULTIPLE ORIGIN-DESTINATION PAIRS

We conclude this section by briefly considering the generalization of the network routing model to multiple origin-destination pairs. Suppose that there are K such pairs (some of them having possibly the same origin or destination). Assume that origin-destination pair j has total traffic r_j. Let us denote the set of paths for origin-destination pair j by $\mathcal{P}_j$, and now $\mathcal{P} = \cup_j \mathcal{P}_j$. Then the socially optimal routing pattern is a solution to

$$\text{minimize} \quad \sum_{i \in E} x_i l_i(x_i)$$
$$\text{subject to} \quad \sum_{\{p \in \mathcal{P} | i \in p\}} x_p = x_i, \ i \in E,$$
$$\sum_{p \in \mathcal{P}_j} x_p = r_j, \ j = 1, \ldots, k, \text{ and } x_p \geq 0 \text{ for all } p \in \mathcal{P}.$$

Turning now to the equilibrium behavior, it can be easily shown that essentially the same characterization theorem for Wardrop equilibrium applies with multiple origin-destination pairs. Formally,

Theorem 3.5 *A feasible routing pattern* $\mathbf{x}^{WE}$ *is a Wardrop equilibrium if and only if it is a solution to*

$$\text{minimize} \quad \sum_{i \in E} \int_0^{x_i} l_i(z) \, dz$$
$$\text{subject to} \quad \sum_{\{p \in \mathcal{P} | i \in p\}} x_p = x_i, \ i \in E,$$
$$\sum_{p \in \mathcal{P}_j} x_p = r_j, \ j = 1, \ldots, k, \text{ and } x_p \geq 0 \text{ for all } p \in \mathcal{P}.$$

Moreover, if each l_i *is strictly increasing, then* $\mathbf{x}^{WE}$ *is uniquely defined.*

Regarding equilibrium efficiency, it turns out that the PoA bounds obtained for single source-destination remains so when allowing for multiple source-destination pairs; see [90] for further details.

3.2 PARTIALLY OPTIMAL ROUTING

3.2.1 BACKGROUND AND MOTIVATION

Since the passage of the Telecommunications Act in 1996, the Internet has undergone a dramatic transformation and experienced increasing decentralization. Today, thousands of network providers cooperate and compete to provide end-to-end network service to billions of users worldwide. While end-users care about the performance across the entire network, individual network providers optimize their own objectives. The Internet's architecture provides no guarantees that provider incentives will be aligned with end-user objectives.

The emergence of *overlay routing* over the past decade has further highlighted the potentially conflicting objectives of the service provider and the end-users. In overlay routing, end-user software (such as peer-to-peer file-sharing software) makes route selection decisions on the basis of the best end-to-end performance available at any given time, while administrative domains control the routing of traffic within their own (sub)networks. Network operators use traffic engineering to optimize performance, and also to react to the global routing decisions of overlay networks (e.g., [85]).

These considerations make it clear that the study of routing patterns and performance in large-scale communication networks requires an analysis of *partially optimal routing*, where end-to-end route selection is selfish and responds to aggregate route latency, but network providers redirect traffic within their own networks to achieve minimum intradomain total latency.

We develop and analyze in this section a model of partially optimal routing, combining selfish across-domain routing and *traffic engineering* by service providers within their administrative domains.

We consider routing flows between multiple source-destination pairs through a network. As in the previous section, each link is endowed by a *latency function* describing the congestion level (e.g., delay or probability of packet loss) as a function of the total flow passing through the link. Each source-destination pair in the network has a fixed amount of flow, and flows follow the minimum delay route among the available paths as captured by the notion of *Wardrop equilibrium*. Our innovation is to allow subsets of the links in the network ("subnetworks") to be independently owned and operated by different providers, and consider the possibility that these providers engage in traffic engineering and route traffic to minimize the total (or average) latency within their subnetworks. Source-destination pairs sending traffic across subnetworks perceive only the *effective latency* resulting from the traffic engineering of the service providers. The resulting equilibrium, which we call a *partially optimal routing* (POR) equilibrium, is a Wardrop equilibrium according to the effective latencies seen by the source-destination pairs. This model provides a stylized description of the practice of traffic engineering in the Internet. The content of this section is based on [3].

3.2.2   THE MODEL

The network model, the definition of the socially optimal solution and the definition of the Wardrop equilibrium remain the same as in Section 3.1. We now assume that a single network provider controls a subnetwork with unique entry and exit points; within this domain, the provider optimizes performance of traffic flow. Formally, we assume there is a disjoint collection of directed subgraphs (subnetworks) inside of G. Within a subnetwork $G_0 = (V_0, A_0)$, a service provider *optimally* routes all incoming traffic. Let $s_0 \in V_0$ denote the unique entry point to G_0, and let $t_0 \in V_0$ denote the unique exit point from G_0. Let P_0 denote the set of available paths from s_0 to t_0 using the edges in A_0. We make the assumption that every path in P passing through any node in V_0 must contain a path in P_0 from s_0 to t_0; this is consistent with our assumption that G_0 is an independent autonomous system, with a unique entry and exit point. We call $R_0 = (V_0, A_0, P_0, s_0, t_0)$ a *subnetwork* of G, and with a slight abuse of notation, we say that $R_0 \subset R$.

Given an incoming amount of flow X_0, the network provider chooses a routing of flow to solve the following optimization problem to minimize total (or average) latency:

$$\text{minimize} \quad \sum_{j \in A_0} x_j l_j(x_j) \tag{3.3}$$

$$\text{subject to} \quad \sum_{p \in P_0 : j \in p} y_p = x_j, \quad j \in A_0;$$

$$\sum_{p \in P_0} y_p = X_0;$$

$$y_p \geq 0, \quad p \in P_0.$$

In this optimization problem, the subnetwork owner sees an incoming traffic amount X_0, and chooses the optimal routing of this flow through the subnetwork. This is a formal abstraction of the process of *traffic engineering* carried out by many network providers to optimize intradomain performance.

Let $L(X_0)$ denote the optimal value of the preceding optimization problem. We define $l_0(X_0) = L(X_0)/X_0$ as the *effective latency* of partially optimal routing in the subnetwork R_0, with flow $X_0 > 0$. If traffic in the entire network G routes selfishly, while traffic is optimally routed within G_0, then replacing G_0 by a single link with latency l_0 will leave the Wardrop equilibrium flow unchanged elsewhere in G.

We have the following simple lemma that provides basic properties of l_0 and L.

Lemma 3.6 *Assume that every latency function, l_j, is a strictly increasing, nonnegative, and continuous function. Then:*

(a) *The effective latency $l_0(X_0)$ is a strictly increasing function of $X_0 > 0$.*

(b) *Assume further that each l_j is a convex function. The total cost $L(X_0)$ is a convex function of X_0.*

In light of the preceding lemma, we can extend the definition of l_0 so that $l_0(0) = \lim_{x_0 \downarrow 0} l_0(x_0)$; the preceding limit is well defined since l_0 is strictly increasing.

To define the overall network performance under partially optimal routing, first suppose that there is a single independently-operated subnetwork. Given a routing instance $R = (V, A, P, s, t, X, l)$, and a subnetwork $R_0 = (V_0, A_0, P_0, s_0, t_0)$ defined as above, we define a new routing instance $R' = (V', A', P', s, t, X, l')$ as follows:

$$V' = (V \setminus V_0) \bigcup \{s_0, t_0\};$$

$$A' = (A \setminus A_0) \bigcup \{(s_0, t_0)\};$$

P' corresponds to all paths in P, where any subpath in P_0 is replaced by the link (s_0, t_0); and l' consists of latency functions l_j for all edges in $A \setminus A_0$, and latency l_0 for the edge (s_0, t_0). Thus, R'

is the routing instance R with the subgraph G_0 replaced by a single link with latency l_0; we call R' the *equivalent POR instance* for R with respect to R_0. The overall network flow in R with partially optimal routing in R_0, $x^{POR}(R, R_0)$, is defined to be the Wardrop equilibrium flow in the routing instance R':

$$x^{POR}(R, R_0) = x^{WE}(R').$$

In other words, it is equilibrium with traffic routed selfishly given the effective latency l_0 of the subnetwork R_0. Note also that this formulation leaves undefined the exact flow in the subnetwork R_0; this is to be expected since problem (3.3) may not have a unique solution.

The total latency cost of the equivalent POR instance for R with respect to R_0 is given by

$$C(x^{POR}(R, R_0)) = \sum_{j \in A'} x_j^{POR}(R, R_0) l_j (x_j^{POR}(R, R_0)).$$

3.2.3   EFFICIENCY OF PARTIALLY OPTIMAL ROUTING

We first consider the effect of optimal routing within subnetworks on the performance of the overall network. One might conjecture that optimally routing traffic within subnetworks should improve the overall performance. The following example shows that this need not be the case.

Example 3.7   Consider the network $G = (V, A)$ with source and destination nodes $s, t \in V$ illustrated in Figure 3.4(a). Let $R = (V, A, P, s, t, 1, 1)$ be the corresponding routing instance, i.e., one unit of flow is to be routed over this network. The subnetwork G_0 consists of the two parallel links in the middle, links 5 and 6, with latency functions

$$l_5(x_5) = 0.31, \qquad l_6(x_6) = 0.4\, x_6.$$

The latency functions for the remaining links in the network are given by

$$l_1(x_1) = x_1, \qquad l_2(x_2) = 3.25,$$

$$l_3(x_3) = 1.25, \qquad l_4(x_4) = 3x_4.$$

Assume first that the flow through the subnetwork G_0 is routed selfishly, i.e., according to the Wardrop equilibrium. Given a total flow X_0 through the subnetwork G_0, the effective Wardrop latency can be defined as

$$\tilde{l}_0(X_0) = \frac{1}{X_0} C(x^{WE}(R_0)), \tag{3.4}$$

where R_0 is the routing instance corresponding to the subnetwork G_0 with total flow X_0. The effective Wardrop latency for this example is given by

$$\tilde{l}_0(X_0) = \min\{0.31, 0.4X_0\}.$$

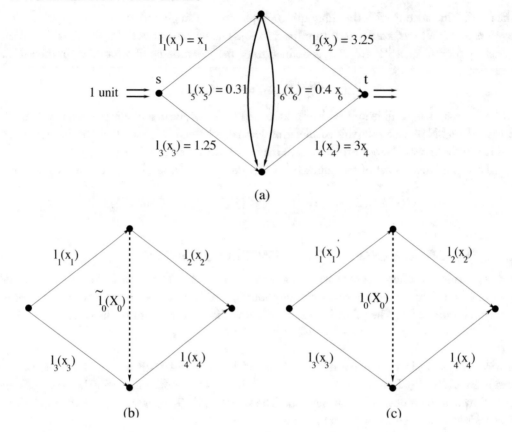

(a)

(b)   (c)

Figure 3.4: A network for which POR leads to a worse performance relative to selfish routing. Figures (b) and (c) illustrate representing the subnetwork with a single link with Wardrop effective latency $\tilde{l}_0(X_0)$ and optimal effective latency $l_0(X_0)$, respectively.

Substituting the subnetwork with a single link with latency function $\tilde{l}_0$ yields the network in Figure 3.4(b). It can be seen that selfish routing over the network of Figure 3.4(b) leads to the link flows $x_1^{WE} = 0.94$ and $X_0^{WE} = 0.92$, with a total cost of $C(x^{WE}(R)) = 4.19$. It is clear that this flow configuration arises from a Wardrop equilibrium in the original network.

Assume next that the flow through the subnetwork G_0 is routed optimally, i.e., as the optimal solution of problem (3.3) for the routing instance corresponding to G_0. Given a total flow X_0 through the subnetwork G_0, the effective latency of optimal routing within the subnetwork G_0 can be defined as

$$l_0(X_0) = \frac{L(X_0)}{X_0},$$

where $L(X_0)$ is the optimal value of problem (3.3). The effective optimal routing latency for this example is given by

$$l_0(X_0) = \begin{cases} 0.4X_0, & \text{if } 0 \leq X_0 \leq 0.3875; \\ 0.31 - \frac{0.0961}{1.6X_0}, & \text{if } X_0 \geq 0.3875. \end{cases}$$

Substituting the subnetwork with a single link with latency function l_0 yields the network in Figure 3.4(c). Note that selfish routing over this network leads to the partially optimal routing (POR) equilibrium. It can be seen that at the POR equilibrium, the link flows are given by $x_1^{POR} = 1$ and $X_0^{POR} = 1$, with a total cost of $C(x^{POR}(R)) = 4.25$, which is strictly greater than $C(x^{WE}(R))$.

In this section, we quantify the inefficiency of partially optimal routing. Our metric of efficiency is the *ratio* of the total cost at the social optimum to the total cost at the partially optimal routing solution, $C(x^{SO})/C(x^{POR})$. Throughout, we assume that all independently-operated subnetworks can be represented as subgraphs with unique entry and exit points.

We will establish two main theorems. The first provides a tight bound on the loss of efficiency when all latency functions are affine; and the second provides a tight bound on the loss of efficiency when all latency functions are polynomials of bounded degree.

We start with a simple result that compares the worst-case efficiency loss of partially optimal routing with that of selfish routing. These relations will be useful in finding tight bounds on the efficiency loss of partially optimal routing. Recall that $\mathcal{R}^{conv}$, $\mathcal{R}^{aff}$, and $\mathcal{R}^{conc}$ denote the class of all routing instances where latency functions are convex, affine, and concave, respectively.

Proposition 3.8

(a) For all $\mathcal{R}' \in \{\mathcal{R}^{conv}, \mathcal{R}^{aff}, \mathcal{R}^{conc}\}$, we have

$$\inf_{\substack{R \in \mathcal{R}' \\ R_0 \subset R}} \frac{C(x^{SO}(R))}{C(x^{POR}(R, R_0))} \leq \inf_{R \in \mathcal{R}'} \frac{C(x^{SO}(R))}{C(x^{WE}(R))}. \tag{3.5}$$

(b)

$$\inf_{\substack{R \in \mathcal{R} \\ R_0 \subset R}} \frac{C(x^{SO}(R))}{C(x^{POR}(R, R_0))} = \inf_{R \in \mathcal{R}} \frac{C(x^{SO}(R))}{C(x^{WE}(R))}.$$

(c)

$$\inf_{\substack{R \in \mathcal{R}^{aff} \\ R_0 \subset R}} \frac{C(x^{SO}(R))}{C(x^{POR}(R, R_0))} \geq \inf_{R \in \mathcal{R}^{conc}} \frac{C(x^{SO}(R))}{C(x^{WE}(R))}.$$

Our main theorem in this section is an extension of the results in Theorem 3.4 to the setting of partially optimal routing.

Theorem 3.9

(a)

$$\inf_{\substack{R \in \mathcal{R}^{conv} \\ R_0 \subset R}} \frac{C(x^{SO}(R))}{C(x^{POR}(R, R_0))} = 0.$$

(b) *Consider a routing instance $R = (V, A, P, s, t, X, l)$ where l_j is an affine latency function for all $j \in A$; and a subnetwork R_0 of R. Then:*

$$\frac{C(x^{SO}(R))}{C(x^{POR}(R, R_0))} \geq \frac{3}{4}.$$

Furthermore, the bound above is tight.

Proof. Part (a) of the theorem is an immediate corollary of Proposition 3.8(a) (for $\mathcal{R}' = \mathcal{R}^{conv}$) and Theorem 3.4(a).

The remainder of the proof establishes part (b) of the theorem by proving two lemmas. The first provides a tight bound of $3/4$ on the ratio of the optimal routing cost to the selfish routing cost for routing instances in which the latency function of each link is a *concave* function. This lemma is relevant because when all latency functions are affine, the effective latency of any subnetwork under partially optimal routing is concave, as shown in the second lemma.

The proof of the following lemma uses a geometric argument that was used in [36]. This result also follows from the analysis in [35]. Here, we provide an alternative proof, which will be useful in our subsequent analysis.

Lemma 3.10   *Let $R \in \mathcal{R}^{conc}$ be a routing instance where all latency functions are concave. Then,*

$$\frac{C(x^{SO}(R))}{C(x^{WE}(R))} \geq \frac{3}{4}.$$

Furthermore, this bound is tight.

Proof of Lemma. Consider a routing instance $R \in \mathcal{R}^{conc}$, with $R = (V, A, P, s, t, X, l)$. Let x^{WE} be the flow configuration at a Wardrop equilibrium. Recall that x^{WE} is a Wardrop equilibrium if and only if it satisfies

$$\sum_{j \in A} l_j(x_j^{WE})(x_j^{WE} - x_j) \leq 0, \tag{3.6}$$

for all feasible solutions x for the same routing instance (see, e.g., [39]).

By Eq. (3.6), for all feasible solutions $\mathbf{x}$ of Problem (3.2), we have

$$C(\mathbf{x}^{WE}) = \sum_{j \in A} x_j^{WE} l_j(x_j^{WE}) \tag{3.7}$$

$$\leq \sum_{j \in A} x_j l_j(x_j^{WE}) \tag{3.8}$$

$$= \sum_{j \in A} x_j l_j(x_j) + \sum_{j \in A} x_j(l_j(x_j^{WE}) - l_j(x_j)).$$

We next show that for all $j \in A$, and all feasible solutions $\mathbf{x}$ of Problem (3.2), we have

$$x_j(l_j(x_j^{WE}) - l_j(x_j)) \leq \frac{1}{4} x_j^{WE} l_j(x_j^{WE}). \tag{3.9}$$

If $x_j \geq x_j^{WE}$, then since l_j is nondecreasing, we have $l_j(x_j^{WE}) \leq l_j(x_j)$, establishing the desired relation (3.9). Assume next that $x_j < x_j^{WE}$. The term $x_j(l_j(x_j^{WE}) - l_j(x_j))$ is equal to the area of the shaded rectangle in Figure 1. Consider the triangle formed by the three points

$$(0, l_j(x_j^{WE})), \ (0, l_j(x_j) - l_j'(x_j)x_j),$$

$$\left(\frac{l_j(x_j^{WE}) - l_j(x_j) + l_j'(x_j)x_j}{l_j'(x_j)}, l_j(x_j^{WE}) \right).$$

Denote this triangle by T. It can be seen that

$$x_j(l_j(x_j^{WE}) - l_j(x_j)) \leq \frac{1}{2} \text{Area}(T).$$

By the concavity of l_j, we further have

$$\text{Area}(T) \leq \int_0^{x_j^{WE}} \int_{l_j(x)}^{l_j(x_j^{WE})} dy\,dx \leq \frac{x_j^{WE} l_j(x_j^{WE})}{2},$$

i.e., the area of triangle T is less than or equal to the area between the curves $y = l_j(x)$ and $y = l_j(x_j^{WE})$ in the interval $x \in [0, x_j^{WE}]$, which in turn is less than or equal to half of the area $x_j^{WE} l_j(x_j^{WE})$. Combining the preceding two relations, we obtain Eq. (3.9), which implies

$$\sum_{j \in A} x_j(l_j(x_j^{WE}) - l_j(x_j)) \leq \frac{1}{4} \sum_{j \in A} x_j^{WE} l_j(x_j^{WE}) = \frac{1}{4} C(\mathbf{x}^{WE}).$$

Combining with Eq. (3.8), we see that for all feasible solutions $\mathbf{x}$ of Problem (3.2), we have

$$\frac{3}{4} C(\mathbf{x}^{WE}) \leq \sum_{j \in A} x_j l_j(x_j).$$

Since the socially optimal flow configuration x^{SO} is a feasible solution for Problem (3.2), we obtain the desired result. ∎

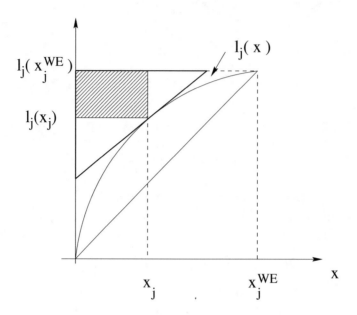

Figure 3.5: Illustration of the proof of Proposition 3.10.

The following lemma, which establishes that the effective latency l_0 of a subnetwork under partially optimal routing is concave when the latency functions are affine, completes the proof of part (b) of the theorem.

Lemma 3.11   *Let $R_0 = (V_0, A_0, P_0, s_0, t_0)$ be a subnetwork. Assume that the latency functions of all links in the subnetwork are nonnegative affine functions, i.e., for all $j \in A_0$, $l_j(x_j) = a_j x_j + b_j$, where $a_j \geq 0$ and $b_j \geq 0$. Let $l_0(X_0)$ denote the effective latency of partially optimal routing of X_0 units of flow in the subnetwork R_0. Then $l_0(X_0)$ is a concave function of X_0.*

Proof of Lemma. Since the l_j are affine, for all $X_0 \geq 0$, we have

$$l_0(X_0) = \min_{y_p \geq 0,\ p \in P} \sum_{j \in A_0} \frac{a_j x_j^2}{X_0} + \frac{b_j x_j}{X_0}$$

$$\text{subject to} \quad \sum_{p \in P_0: j \in p} y_p = x_j, \quad j \in A_0;$$

$$\sum_{p \in P_0} y_p = X_0.$$

Using the change of variables $\hat{y}_p = \frac{y_p}{X_0}$ for all $p \in P_0$, and $\hat{x}_j = \frac{x_j}{X_0}$ for all $j \in A_0$ in the preceding optimization problem, we obtain

$$l_0(X_0) = \min_{\hat{y}_p \geq 0, \; p \in P_0} \sum_{j \in A_0} a_j X_0 \hat{x}_j^2 + b_j \hat{x}_j \tag{3.10}$$

$$\text{subject to} \quad \sum_{p \in P_0 : j \in p} \hat{y}_p = \hat{x}_j, \quad j \in A_0;$$

$$\sum_{p \in P_0} \hat{y}_p = 1.$$

Denote the feasible set of problem (3.10) by Y, i.e.,

$$Y = \left\{ y \; \middle| \; y_p \geq 0, \; \forall \, p \in P_0, \; \sum_{p \in P_0} y_p = 1 \right\}.$$

Then by defining $x_j(y) = \sum_{p \in P_0 : j \in p} y_p$, we can write (3.10) equivalently as:

$$l_0(X_0) = \inf_{y \in Y} \left[\left(\sum_{j \in A_0} a_j x_j(y)^2 \right) X_0 + \left(\sum_{j \in A_0} b_j x_j(y) \right) \right].$$

But now observe that $l_0(X_0)$ is the infimum of a collection of affine functions of X_0. By a standard result in convex analysis (see, e.g., [21], Proposition 1.2.4(c)), it follows that $l_0(X_0)$ is concave. ∎

Combining Lemmas 3.10 and 3.11 with Proposition 3.8 completes the proof of part (b) of Theorem 3.9. □

3.2.4 EXTENSIONS

To conclude this section, we next provide a brief description of additional results that were obtained in [3]. We first mention that this paper provides tight bounds on the loss of efficiency when all latency functions are polynomials of bounded degree.

In contrast to the results in Theorem 3.9 that match the corresponding bounds for selfish routing throughout the whole network, when subnetworks have multiple entry-exit points, the performance of partially optimal routing can be arbitrarily bad, even with linear latencies. This result suggests that special care needs to be taken in the regulation of traffic in large-scale networks overlaying selfish source routing together with traffic engineering within subnetworks.

The paper [3] also provides conditions for service providers to prefer to engage in traffic engineering rather than allowing all traffic to route selfishly within their network. The latter is a possibility because selfish routing may discourage entry of further traffic into their subnetwork, reducing total delays within the subnetwork, which may be desirable for the network provider when there are no prices per unit of transmission.

Overall, we believe that the model of partially optimal routing presented in this section is a good approximation to the functioning of large-scale communication networks, such as the Internet.

3.3   CONGESTION AND PROVIDER PRICE COMPETITION

3.3.1   PRICING AND EFFICIENCY WITH CONGESTION EXTERNALITIES

We now construct a model of resource allocation in a network with competing selfish users and profit-maximizing service providers. The central question is whether the equilibrium prices that emerge in such a framework affect system efficiency. The class of models incorporating strategic behavior by service providers introduces new modeling and mathematical challenges. These models translate into game-theoretic competition models with negative congestion externalities,[3] whereby the pricing decision of a service provider affects the level of traffic and thus the extent of congestion in other parts of the network. Nevertheless, tractable analysis of pricing decisions and routing patterns are possible under many network topologies.

Models incorporating for-profit service providers have been previously investigated in [16], [17], and [4]. Here, we develop a general framework for the analysis of price competition among providers in a congested (and potentially capacitated) network building on [5] and [6]. We will see that despite its conceptual simplicity, this framework has rich implications. We illustrate some of these, for example, by showing the counterintuitive result that increasing competition among providers can reduce efficiency, which is different from the results of the most common models of competition in economics. Most importantly, we also show that it is possible to quantify the extent to which prices set by competing service providers affect system efficiency. While generally service provider competition does not lead to an equilibrium replicating the system optimum, the extent of inefficiency resulting from price competition among service providers can often be bounded.

We start with a simple example which shows the efficiency implications of competition between two for-profit service providers.

Example 3.12   One unit of traffic will travel from an origin to a destination using either route 1 or route 2 (cf. Figure 3.6). The latency functions of the links, which represent the delay costs as a function of the total link flow, are given by

$$l_1(x) = \frac{x^2}{3}, \qquad l_2(x) = \frac{2}{3}x.$$

It is straightforward to see that the efficient allocation [i.e., one that minimizes the total delay cost $\sum_i l_i(x_i)x_i$] is $x_1^S = 2/3$ and $x_2^S = 1/3$, while the (Wardrop) equilibrium allocation that equates delay on the two paths is $x_1^{WE} \approx .73 > x_1^S$ and $x_2^{WE} \approx .27 < x_2^S$. The source of the inefficiency is that each unit of traffic does not internalize the greater increase in delay from travel on route 1, so there is too much use of this route relative to the efficient allocation.

[3]An externality arises when the actions of the player in a game affects the payoff of other players.

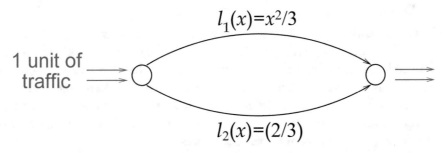

Figure 3.6: A two link network with congestion-dependent latency functions.

Now consider a monopolist controlling both routes and setting prices for travel to maximize its profits. We show below that in this case, the monopolist will set a price including a markup, which exactly internalizes the congestion externality. In other words, this markup is equivalent to the Pigovian tax that a social planner would set in order to induce decentralized traffic to choose the efficient allocation. Consequently, in this simple example, monopoly prices will be $p_1^{ME} = (2/3)^3 + k$ and $p_2^{ME} = (2/3^2) + k$, for some constant k. The resulting traffic in the Wardrop equilibrium will be identical to the efficient allocation, i.e., $x_1^{ME} = 2/3$ and $x_2^{ME} = 1/3$.

Finally, consider a duopoly situation, where each route is controlled by a different profit-maximizing provider. In this case, it can be shown that equilibrium prices will take the form $p_i^{OE} = x_i^{OE} \left(l_1' + l_2' \right)$ [see Eq. (3.16) in Section 3.3.4], or more specifically, $p_1^{OE} \approx 0.61$ and $p_2^{OE} \approx 0.44$. The resulting equilibrium traffic is $x_1^{OE} \approx .58 < x_1^S$ and $x_2^{OE} \approx .42 > x_2^S$, which also differs from the efficient allocation. It is noteworthy that although the duopoly equilibrium is inefficient relative to the monopoly equilibrium, in the monopoly equilibrium k is chosen such that all of the consumer surplus is captured by the monopolist, while in the oligopoly equilibrium users may have positive consumer surplus.[4]

The intuition for the inefficiency of the duopoly relative to the monopoly is related to a new source of (differential) monopoly power for each duopolist, which they exploit by distorting the pattern of traffic: when provider 1, controlling route 1, charges a higher price, it realizes that this will push some traffic from route 1 to route 2, raising congestion on route 2. But this makes the traffic using route 1 become more "locked-in," because their outside option, travel on the route 2, has become worse. As a result, the optimal price that each duopolist charges will include an additional markup over the Pigovian markup. Since the two markups are generally different, they will distort the pattern of traffic away from the efficient allocation.

---

[4]Consumer surplus is the difference between users' willingness to pay (reservation price) and effective costs, $p_i + l_i(x_i)$, and is thus different from the social surplus (which is the difference between users' willingness to pay and latency cost, $l_i(x_i)$, thus also takes into account producer surplus/profits).

3.3.2 MODEL

We consider a network with I parallel links. Let $\mathcal{I} = \{1, \ldots, I\}$ denote the set of links. Let x_i denote the total flow on link i, and $x = [x_1, \ldots, x_I]$ denote the vector of link flows. Each link in the network has a flow-dependent latency function $l_i(x_i)$, which measures the delay as a function of the total flow on link i. We assume that the latency function l_i is convex, nondecreasing, and continuously differentiable. The analysis can be extended to the case when the links are capacity-constrained, see [5]. We also assume that $l_i(0) = 0$ for all i[5]. We denote the price per unit flow (bandwidth) of link i by p_i. Let $p = [p_1, \ldots, p_I]$ denote the vector of prices.

We are interested in the problem of routing d units of flow across the I links. We assume that this is the aggregate flow of many "small" users and thus adopt the Wardrop's principle in characterizing the flow distribution in the network; i.e., the flows are routed along paths with minimum effective cost, defined as the sum of the latency at the given flow and the price of that path. We also assume that the users have a homogeneous *reservation utility* R and decide not to send their flow if the effective cost exceeds the reservation utility.

More formally, for a given price vector $p \geq 0$, a vector $x^{WE} \in \mathbb{R}^I_+$ is a *Wardrop equilibrium* (WE) if

$$l_i(x_i^{WE}) + p_i = \min_j \{l_j(x_j^{WE}) + p_j\}, \qquad \forall i \text{ with } x_i^{WE} > 0, \qquad (3.11)$$

$$l_i(x_i^{WE}) + p_i \leq R, \qquad \forall i \text{ with } x_i^{WE} > 0,$$

$$\sum_{i \in \mathcal{I}} x_i^{WE} \leq d,$$

with $\sum_{i \in \mathcal{I}} x_i^{WE} = d$ if $\min_j \{l_j(x_j^{WE}) + p_j\} < R$. We denote the set of WE at a given p by $W(p)$.[6]

We next define the social problem and the social optimum, which is the routing (flow allocation) that would be chosen by a planner that has full information and full control over the network. A flow vector x^S is a *social optimum* if it is an optimal solution of the *social problem*

$$\max_{\substack{x \geq 0 \\ \sum_{i \in \mathcal{I}} x_i \leq d}} \sum_{i \in \mathcal{I}} \left(R - l_i(x_i)\right) x_i. \qquad (3.12)$$

Hence, the social optimum is the flow allocation that maximizes the social surplus, i.e., the difference between users' willingness to pay and total latency. For two links, let x^S be a social optimum with $x_i^S > 0$ for $i = 1, 2$. Then it follows from first order optimality conditions that

$$l_1(x_1^S) + x_1^S l_1'(x_1^S) = l_2(x_2^S) + x_2^S l_2'(x_2^S). \qquad (3.13)$$

This implies that the prices $x_i^S l_i'(x_i^S)$, i.e., the marginal congestion prices, can be used to enforce the system optimum [cf. Eq. (3.11)].

---

[5]This assumption is a good approximation to communication networks where queueing delays are more substantial than propagation delays. We will talk about the efficiency implications of relaxing this assumption in different models.

[6]It is possible to account for additional constraints, such as capacity constraints on the links, by using a variational inequality formulation (see [5], [36]).

For a given vector $x \geq 0$, we define the value of the objective function in the social problem,

$$\mathbb{S}(x) = \sum_{i \in \mathcal{I}} (R - l_i(x_i)) \, x_i, \tag{3.14}$$

as the *social surplus*, i.e., the difference between users' willingness to pay and the total latency.

3.3.3   MONOPOLY PRICING AND EQUILIBRIUM

We first assume that a monopolist service provider owns the I links and charges a price of p_i per unit bandwidth on link i. The monopolist sets the prices to maximize his profit given by

$$\Pi(p, x) = \sum_{i \in \mathcal{I}} p_i x_i,$$

where $x \in W(p)$. This defines a two-stage dynamic *pricing-congestion game*, where the monopolist sets prices anticipating the demand of users, and given the prices (i.e., in each subgame), users choose their flow vectors according to the WE. We define a vector $(p^{ME}, x^{ME}) \geq 0$ to be a *Monopoly Equilibrium* (ME) if $x^{ME} \in W(p^{ME})$ and

$$\Pi(p^{ME}, x^{ME}) \geq \Pi(p, x), \qquad \forall \, p \geq 0, \; \forall \, x \in W(p).^7$$

In [5], it was shown that price-setting by a monopolist internalizes the negative externality and achieves efficiency. In particular, a vector x is the flow vector at an ME if and only if it is a social optimum. This result was extended to a model that incorporates a general network topology in [51]. This is a significant departure from the existing performance results of selfish routing in the literature which assert that the efficiency losses with general latency functions can be arbitrarily bad.

3.3.4   OLIGOPOLY PRICING AND EQUILIBRIUM

We next assume that there are S service providers, denote the set of service providers by $\mathcal{S}$, and assume that each service provider $s \in \mathcal{S}$ owns a different subset $\mathcal{I}_s$ of the links. Service provider s charges a price p_i per unit bandwidth on link $i \in \mathcal{I}_s$. Given the vector of prices of links owned by other service providers, $p_{-s} = [p_i]_{i \notin \mathcal{I}_s}$, the profit of service provider s is

$$\Pi_s(p_s, p_{-s}, x) = \sum_{i \in \mathcal{I}_s} p_i x_i,$$

for $x \in W(p_s, p_{-s})$, where $p_s = [p_i]_{i \in \mathcal{I}_s}$.

The objective of each service provider, like the monopolist in the previous section, is to maximize profits. Because their profits depend on the prices set by other service providers, each service provider forms conjectures about the actions of other service providers, as well as the behavior

---

[7]Our definition of the ME is stronger than the standard subgame perfect Nash equilibrium concept for dynamic games. In [5], we show that the two solution concepts coincide for this game.

of users, which, we assume, they do according to the notion of (subgame perfect) Nash equilibrium. We refer to the game among service providers as the *price competition game*. We define a vector $(p^{OE}, x^{OE}) \geq 0$ to be a (pure strategy) *Oligopoly Equilibrium* (OE) if $x^{OE} \in W\left(p_s^{OE}, p_{-s}^{OE}\right)$ and for all $s \in \mathcal{S}$,

$$\Pi_s(p_s^{OE}, p_{-s}^{OE}, x^{OE}) \geq \Pi_s(p_s, p_{-s}^{OE}, x), \qquad \forall \, p_s \geq 0, \ \forall \, x \in W(p_s, p_{-s}^{OE}). \tag{3.15}$$

We refer to p^{OE} as the *OE price*.

Analysis of the optimality conditions for the oligopoly problem [cf. (3.15)] allows us to characterize the OE prices (see [5]). In particular, let (p^{OE}, x^{OE}) be an OE such that $p_i^{OE} x_i^{OE} > 0$ for some $i \in \mathcal{I}$. Then, for all $s \in \mathcal{S}$ and $i \in \mathcal{I}_s$,

$$p_i^{OE} = \begin{cases} x_i^{OE} l_i'(x_i^{OE}), & \text{if } l_j'(x_j^{OE}) = 0 \text{ for some } j \notin \mathcal{I}_s, \\[2ex] \min\left\{ R - l_i(x_i^{OE}) \ , \ x_i^{OE} l_i'(x_i^{OE}) + \dfrac{\sum_{j \in \mathcal{I}_s} x_j^{OE}}{\sum_{j \notin \mathcal{I}_s} \frac{1}{l_j'(x_j^{OE})}} \right\}, & \text{otherwise.} \end{cases}$$

The preceding characterization implies that in the two link case with minimum effective cost less than R, the OE prices satisfy

$$p_i^{OE} = x_i^{OE}(l_1'(x_1^{OE}) + l_2'(x_2^{OE})) \tag{3.16}$$

as claimed before. Intuitively, the price charged by an oligopolist consists of two terms: the first, $x_i^{OE} l_i'(x_i^{OE})$, is equal to the marginal congestion price that a social planner would set [cf. Eq. (3.13)] because the service provider internalizes the further congestion caused by additional traffic. The second, $x_i^{OE} l_j'(x_j^{OE})$, reflects the markup that each service provider can charge users because of the negative congestion externality (as users leave its network, they increase congestion in the competitor network).

3.3.5   EFFICIENCY ANALYSIS

We investigate the efficiency properties of price competition games that have pure strategy equilibria [8]. Given a price competition game with latency functions $\{l_i\}_{i \in \mathcal{I}}$, we define the efficiency metric at some oligopoly equilibrium flow x^{OE} as the ratio of the social surplus in the oligopoly equilibrium to the surplus in the social optimum [cf. Eq. 3.14 for the definition of the social surplus], i.e., the efficiency metric is given by

$$r_I(\{l_i\}, x^{OE}) = \frac{\mathbb{S}(x^{OE})}{\mathbb{S}(x^S)}, \tag{3.17}$$

where x^S is a social optimum given the latency functions $\{l_i\}_{i \in \mathcal{I}}$ and R is the reservation utility. In other words, the efficiency metric is the ratio of the social surplus in an equilibrium relative to the surplus in the social optimum. As in the previous sections, we are interested in the worst-case

---

[8]This set includes, but is substantially larger than, games with linear latency functions, see [6].

performance (or Price of Anarchy) of an oligopoly equilibrium, so we look for a lower bound on $r_I(\{l_i\}, x^{OE})$ over all price competition games and all oligopoly equilibria.

We next give an example of an I link network which has positive flows on all links at the OE and an efficiency metric of 5/6.

Example 3.13   Consider an I link network where each link is owned by a different provider. Let the total flow be $d = 1$ and the reservation utility be $R = 1$. The latency functions are given by

$$l_1(x) = 0, \qquad l_i(x) = \frac{3}{2}(I - 1)x, \quad i = 2, \ldots, I.$$

The unique social optimum for this example is $x^S = [1, 0, \ldots, 0]$. It can be seen that the flow allocation at the unique OE is $x^{OE} = \left[\frac{2}{3}, \frac{1}{3(I-1)}, \ldots, \frac{1}{3(I-1)}\right]$. Hence, the efficiency metric for this example is $r_I(\{l_i\}, x^{OE}) = \frac{5}{6}$.

The next theorem establishes the main efficiency result.

Theorem 3.14   *Consider a general parallel link network with $I \geq 2$ links and S service providers, where provider s owns a set of links $\mathcal{I}_s \subset \mathcal{I}$. Then, for all price competition games with pure strategy OE flow x^{OE}, we have*

$$r_I(\{l_i\}, x^{OE}) \geq \frac{5}{6},$$

and the bound is tight.

The idea behind the proof of the theorem is to lower bound the infinite dimensional optimization problem associated with the definition of the Price of Anarchy by a finite dimensional problem. Then, one can use the special structure of parallel links to analytically solve the finite-dimensional optimization problem. Further details can be found in [5].

A notable feature of Example 3.13 and this theorem is that the (tight) lower bound on inefficiency is independent of the number of links I and how these links are distributed across different oligopolists (i.e., of market structure). Thus, arbitrarily large networks can feature as much inefficiency as small networks.[9]

3.3.6   EXTENSIONS

In this subsection, we extend the preceding analysis in two directions: First, we consider elastic traffic, which models applications that are tolerant of delay and can take advantage of even the minimal amounts of bandwidth (e.g., e-mail). We then focus on more general network topologies.

Elastic Traffic.   To model elastic traffic, we assume that user preferences can be represented by an increasing, concave, and twice continuously differentiable *aggregate* utility function $u\left(\sum_{i \in \mathcal{I}} x_i\right)$,

---

[9]This result superficially contrasts with theorems in the economics literature that large oligopolistic markets approach competitive behavior. These theorems do not consider arbitrary large markets but replicas of a given market structure.

which represents the amount of utility gained from sending a total amount of flow $\sum_{i \in \mathcal{I}} x_i$ through the network.

We assume that at a price vector, the amount of flow and the distribution of flow across the links is given by the Wardrop equilibrium (cf. Def. 3.2). In particular, for a given price vector $p \geq 0$, a vector $x^* \in \mathbb{R}_+^I$ is a Wardrop equilibrium if

$$
\begin{aligned}
l_i(x_i^*) + p_i &= u'\left(\sum_{j \in \mathcal{I}} x_j^*\right), & \forall\, i \text{ with } x_i^* > 0, \\
l_i(x_i^*) + p_i &\geq u'\left(\sum_{j \in \mathcal{I}} x_j^*\right), & \forall\, i \in \mathcal{I}.
\end{aligned}
$$

We define the social optimum and the efficiency metric as in Eqs. (3.12) and (3.17), replacing $R \sum_{i \in \mathcal{I}} x_i$ (i.e., users' willingness to pay) by $u\left(\sum_{i \in \mathcal{I}} x_i\right)$.

It can be shown that for elastic traffic with a general concave utility function, the efficiency metric can be arbitrarily close to 0 (see [82]). The two-stage game with multiple service providers and elastic traffic with a single user class was first analyzed by Hayrapetyan, Tardos, and Wexler [49]. Using an additional assumption on the utility function (i.e., the utility function has a concave first derivative), their analysis provides non-tight bounds on the efficiency loss[10]. Using mathematical tools similar to the analysis in [5], the recent work [82] provides a tight bound on the efficiency loss of this game, as established in the following theorem.

Theorem 3.15   *Consider a parallel link network with $I \geq 1$ links, where each link is owned by a different provider. Assume that the derivative of the utility function, u' is a concave function. Then, for all price competition games with elastic traffic and pure strategy OE flow x^{OE}, we have*

$$
r_I(u, \{l_i\}, x^{OE}) \geq \frac{2}{3},
$$

and the bound is tight.

Parallel-Serial Topologies. Most communication networks cannot be represented by parallel link topologies, however. A given source-destination pair will typically transmit through multiple interconnected subnetworks (or links), potentially operated by different service providers. Existing results on the parallel-link topology do not address how the cooperation and competition between service providers will impact on efficiency in such general networks.

Here, we take a step in this direction by considering the simplest network topology that allows for serial interconnection of multiple links/subnetworks, which is the parallel-serial topology (see Figure 3.7). It was shown in [6] that the efficiency losses resulting from competition are considerably higher with this topology. When a particular provider charges a higher price, it creates

---

[10]For example, they provide the non-tight bound of 1/5.064 in general, and the bound of 1/3.125 for the case when latency without congestion is zero.

a negative externality on other providers along the same path, because this higher price reduces the transmission that all the providers along this path receive. This is the equivalent of the *double marginalization* problem in economic models with multiple monopolies and is the source of the significant degradation in the efficiency performance of the network.

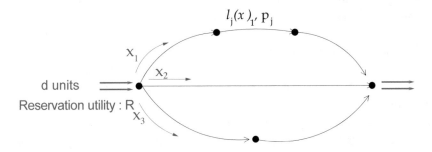

Figure 3.7: A parallel-serial network topology.

In its most extreme form, the double marginalization problem leads to a type of "coordination failure," whereby all providers, expecting others to charge high prices, also charge prohibitively high prices, effectively killing all data transmission on a given path. We may expect such a pathological situation not to arise since firms should not coordinate on such an equilibrium (especially when other equilibria exist). For this reason, we focus on a stronger concept of equilibrium introduced by Harsanyi, the *strict equilibrium*. In strict OE, each service provider must play a strict best response to the pricing strategies of other service providers. We also focus our attention on equilibria in which all traffic is transmitted (otherwise, it can be shown that the double marginalization problem may cause entirely shutting down transmission, resulting in arbitrarily low efficiency, see [6]).

The next theorem establishes the main efficiency result for this topology.

Theorem 3.16   *Consider a general $I \geq 2$ path network, with serial links on each path, where each link is owned by a different provider. Then, for all price competition games with strict OE flow x^{OE}, we have*

$$r_I(x^{OE}) \geq \frac{1}{2},$$

and the bound is tight.

Despite this positive result, it was shown in [6] that when the assumption $l_i(0) = 0$ is relaxed, the efficiency loss of strict OE relative to the social optimum can be arbitrarily large. This suggests that unregulated competition in general communication networks may have considerable costs in terms of the efficiency of resource allocation and certain types of regulation may be necessary to make sure that service provider competition does not lead to significant degradation of network performance.

3.4   CONCLUDING REMARKS

This chapter has focused on basic models of network routing in which users are selfish. We have studied the extent to which selfish behavior affects network performance. In addition to the basic routing game which involves only the end-users themselves, we have addressed networking scenarios in which network providers play an important role. In particular, we have considered communication networks in which service providers engage in traffic engineering, as well as networks where the service providers are profit maximizers.

We conclude this chapter by listing several venues that have not been covered herein, which have been active research areas within the general framework of wireline networks games.

User heterogeneity and service differentiation. The models in this chapter assumed a single user-class in the sense that users are indistinguishable regarding the performance they obtain while using the network, and also regarding their preferences and tradeoffs (e.g., money vs. latency tradeoff). In practice, the user population is often heterogenous, and multi-class user models need to be investigated. From the network side, user heterogeneity requires proper adjustments. A "one-service-class per all" might fail short to accommodate the needs of diverse user types. Consequently, several service classes with different characteristics often need to be offered. Service differentiation brings in a clear need for offering incentives to users to encourage them to choose the service appropriate for their needs, hence preventing over utilization of network resources. Pricing mechanisms provide an efficient way to ensure QoS guarantees and regulate system usage. One of the key debates in network pricing area is whether charges should be based on fixed access prices or usage-based prices. While usage-based pricing has the potential to fulfill at least partially the role of a congestion control mechanism, there were criticisms in view of the apparent disadvantages of billing overheads and the resulting uncertainties in networking expenses (see [42]). A variety of pricing mechanisms have been proposed over the last decade, ranging from simple price differentiation (e.g., Odlyzko Paris Metro Pricing proposal [80]) to auction-based mechanisms and dynamic pricing schemes; see [38, 42] for an overview of various pricing mechanisms. A related engineering issue of service differentiation with a fixed number of service classes, is what kind of QoS guarantees can be offered and advertised, see, e.g., [62] and references therein.

Models for investment and capacity upgrade decisions. Most of the work in this literature of routing games investigates the efficiency losses resulting from the allocation of users and information flows across different paths or administrative domains in an already established network. Arguably, the more important economic decisions in large-scale communication networks concern the *investments* in the structure of the network and in bandwidth capacity. In fact, the last 20 years have witnessed significant investments in broadband, high-speed and optical networks. It is therefore of great importance to model price *and* capacity competition between service providers and investigate the efficiency properties of the resulting equilibria. See [2] and references therein for recent work in the area.

Two-sided markets. The models considered in this section can be viewed as an abstraction of end-users or consumers who route traffic over the Internet. However, the Internet which offers diverse economic opportunities for different parties, involves other important "players", such as *Content Providers*[11] (CPs). Consumers and Content providers (CP) base their choice of Internet Service Providers (ISPs) not only on prices but also on other features such as speed of access, special add-ons like spam blocking and virus protection. Therefore, the competition between service providers is not only over price but also over quality. In addition, the pricing decisions become more involved, as ISPs compete in prices for both CP's and consumers. From an economic perspective, one can treat these models as a competition between interconnected two-sided-market platforms in the presence of quality choice; accordingly, game-theory is a natural paradigm for the analysis of the complex interaction between the different parties. See [79] and references therein for recent work on the subject.

Atomic players. The routing games considered in the section assume that the user population consists of an infinitesimal number of "small" non-atomic users. Such modeling assumption might not be appropriate for certain networking domains, in which each player may control a non-negligible fraction of the total flow. Examples of such domains include logistic and freight companies that transport goods between different points in the world to serve their clients [34].

In general, atomic routing games are harder to analyze when compared to non-atomic ones: The underlying Nash equilibrium might not be unique (see [23, 81, 86]), and except for very special cases, there is no potential function that can be used to characterize the equilibrium point. Atomic network games thus incorporate many research challenges that still need to be explored.

Stochasticity and queuing effects. The use of latency functions as in the routing games considered in this chapter, abstract away the stochastic elements in a communication network. A significant part of the delay in a communication network link is due to the queueing overhead. It is therefore important to incorporate the queueing effects into the selfish routing models. There has been extensive work on decision making in queueing systems, see [48] for a survey book on the subject.

Additional efficiency measures. In this chapter, we have studied the Price of Anarchy (the worst-case ratio between equilibrium and optimal performance) for different selfish routing models. Naturally, there are other performance measures that can be considered. One such is the Price of Stability [90], which is the worst-case ratio between performance at the *best* equilibrium and the social optimum. However, both these measure provide *worst-case* guarantees; one may rightly ask whether networks leading to worst case performance are likely. In other words, worst-case analysis should be supplemented with statistical analysis (e.g., average and variance of the efficiency loss). A related direction is to rule out, or give less significance to equilibria that are less likely to be the outcome of natural user dynamics, thereby replacing the static PoA measure with dynamic PoA analysis.

---

[11] A content provider is defined as an organization or individual that creates information, educational or entertainment content for the Internet.

CHAPTER 4

Wireless Network Games

The emerging use of wireless technologies (such as WiFi and WiMAX) for data communication has brought to focus novel system characteristics which are of less importance in wireline platforms. Power control and the effect of mobility on network performance are good examples of topics which are prominent in the wireless area. An additional distinctive feature of wireless communications is the possible time variation in the channel quality between the sender and the receiver, an effect known as channel fading [25].

Current wireless networks consist of a relatively large number of users with heterogeneous Quality of Service (QoS) requirements (such as bandwidth, delay, and power). To reduce the management complexity, decentralized control of such networks is often to be preferred to centralized one. This requirement leads to distributed (or at least partially distributed) network domains, in which end-users take autonomous decisions regarding their network usage, based on their individual preferences. This framework is naturally formulated as a non-cooperative game, and has gained much interest in recent literature (e.g., see [59] for a recent collection of papers on game theory in communication systems).

In the context of wireless networks, self-interested user behavior can be harmful, as network resources are often limited, and might be abused by a subset of greedy users. In many cases, an individual user can momentarily improve its Quality of Service (QoS) metrics, such as delay and throughput, by accessing the shared channel more frequently. Aggressiveness of even a single user may lead to a chain reaction, resulting in possible throughput collapse.

The wireless domain is quite complex and incorporates many recent technological advances and diverse modeling features. Naturally, we do not intend to cover all of those in this survey. We do provide below a basic taxonomy for classification of the different aspects that define a wireless-network game.

Network structure. A wireless network in its full generality can be described by a set of transmitters and a set of receivers. Each transmitter may in principle be associated with one or more receivers. A transmission to a particular receiver interferes not only with other transmissions to that receiver, but it may also affect data reception at other receivers; see Figure 4.1 for an illustration.

In the bulk of this section, we will focus on the *uplink* network model, where there exists a single receiver (or base station), with multiple nodes transmitting to it.

Multipath propagation and fading. Unlike wired communications that take place over a relatively stable medium, the wireless transmission medium varies strongly with time. The signal that is obtained at the receiver is, in fact, a superposition of multi-path components, which correspond to different propagation paths. If there are phase shifts between the components (due to different

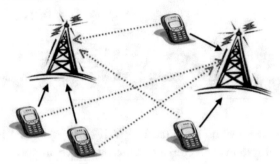

Figure 4.1: A multi-cell wireless network. The transmission power of each mobile affects the throughput of all other mobiles, even if they transmit to different base-stations (interference is illustrated by dotted lines). The higher the power of the mobile the higher its throughput, and the lower the throughput of other mobiles.

interacting objects along the paths), the transmitted signal might not be properly decoded at the receiver. This effect is known as (small-scale) fading. In addition, if there is no line of sight between the transmitter and the receiver, the signal quality obviously diminishes, an effect known as shadowing (or large-scale fading). This effect is often caused by the mobility of the wireless terminals. More generally, the signal's quality at the receiver is inversely proportional to the distance between the transmitter and the receiver, which also varies due to mobility.

The above factors indicate that the channel quality between the transmitter and the receiver is time varying. In this section, we will examine how wireless mobile use the information about the channel quality (referred to as *Channel State Information* – CSI) in order to adjust their transmission parameters. CSI can naturally be exploited for enhanced performance in a centralized setup; however, the consequences of its use under self-interested behavior requires sophisticated game-theoretic analysis.

Reception model. The throughput (i.e., the number of bits that are properly received at the receiver per unit time) of a wireless devise which transmits to some base station depends on numerous factors. First, it depends on the underlying domain (e.g., UMTS cellular system or a wireless LAN network). Within each domain, the throughput could be a function of the transmission rates, the modulation schemes, the packet sizes, the receiver technology, the multi-antenna processing algorithms, and the carrier allocation strategies. In this section, we will not describe the effect of each of the above elements on performance. Nevertheless, we provide below a basic classification of reception models.

1) Capture. As mentioned above, simultaneous transmissions of different mobiles interfere with each other. Perhaps the most basic reception model is that of a *collision channel*, in which a transmission can be successful only if no other user attempts transmission simultaneously. Thus, at each time slot, at most one user can successfully transmit to the base station.

Some wireless LAN receivers are able to decode a single transmission, even if other transmission took place at the same time. The so-called *capture* effect takes place when a single (the strongest) user can be successfully received even in the presence of other simultaneous transmissions, provided that its power dominates the others' transmissions.

A broadly studied capture model is based on the signal to interference plus noise ratio (SINR) (see, e.g., [106] and references therein). Consider an uplink network, and let P_i be user i's transmission power. Denote by h_i the channel gain of user i. Then the SINR for user i is given by

$$\text{SINR}_i(\mathbf{P}) = \frac{h_i P_i}{\sum_{j \neq i} h_j P_j + \sigma_0}. \tag{4.1}$$

A transmission is successful if the SINR is large enough, namely if $\text{SINR}_i > \beta > 1$ (where σ_0 is the ambient noise power).

2) Multi-packet reception. In some wireless systems, such as cellular networks, multiple simultaneous receptions are possible. In the broad sense, these systems rely on spread-spectrum techniques, such as Code Division Multiple Access – CDMA [73]. The throughput of each user is, in general, a function of its SINR (4.1). There are several different ways to model this function, depending on the specific underlying domain. For example, a well-studied model assumes that the user throughput is a logarithmic, concave function of the SINR, namely $r_i(\mathbf{P}) = \log(1 + \text{SINR}_i(\mathbf{P}))$. This expression approximates the case where the transmitter can adjust its coding scheme to obtain rates approaching the Shannon capacity. Another common model for the user throughput is given by $r_i(\mathbf{P}) = R_i f(\text{SINR}_i(\mathbf{P}))$, where R_i is the transmission rate, and $f(\cdot)$ is the packet success probability, which is an increasing function of the SINR (usually sigmoid-shaped, see [68] and references therein).

The choice of the reception model greatly affects the characteristics of the underlying non-cooperative game between the mobile users. Our focus in this section will be on single-packet reception, with emphasis on collision channels. A survey of noncooperative games in multi-packet reception domains can be found in [68].

Preferences and utilities. As mentioned earlier, the choice of utilities in network games reflects certain tradeoffs between QoS measures. In the context of wireless network, the salient measures are throughput and power consumption. Accordingly, the most commonly studied user objectives consist of combinations of these two measures:

- Maximize throughput subject to an average power constraint.

- Minimize the average power consumption subject to a minimum throughput constraint.

- Maximize the ratio between throughput and power consumption. The corresponding utility in this case is referred to as the *bit per joule* utility.

Our focus in this section will be on the objective of minimizing power consumption subject to throughput constraint. In often cases, a pricing term based on resource usage is incorporated into the user utility, in order to regulate the network to the desired operating point. As we shall see in this section, pricing is not always required for social efficiency as the objective of minimizing the power consumption could be self-regulating on its own.

User control and protocol constraints. Depending on the wireless domain, mobile users may in principle be able to self determine various parameters of their transmissions. These include, for example, the timing of the transmission, the transmission rate, the modulation schemes, the packet sizes, and the choice of frequency bands in multi-carrier systems.

If users are given the freedom to control all transmission parameters, then it is expected that their self-interested behavior would lead to undesirable outcomes. Thus, it is often assumed that users may self tune only a subset of the parameters. Moreover, in some cases, the user behavior is regulated by an underlying Medium Access Control (MAC) protocol. A desirable protocol is one which prevents the users from reaching bad equilibria yet still gives each user the right freedom to adjust their transmission parameters based on their heterogenous preferences, thereby leading the network to desirable operating points.

As mentioned above, our focus in this chapter will be on noncooperative mobile interaction in an uplink collision channel. A basic assumption in our models is that the channel quality between each user and the base-station is time-varying, and that each user can monitor its channel quality and adjust its transmission parameters accordingly. The objective of each user is to minimize its energy investment, subject to a minimal throughput constraint. The structure of this chapter is as follows. In Section 4.1, we consider the scenario where each mobile decides on its transmission schedule as a function of its underlying channel state. In Section 4.2, we assume that users can not only adjust their transmission schedule but also control the power level that they employ for every realization of the channel state. Section 4.3 briefly describes related work and highlights several interesting extensions of the above mentioned models. Finally, we provide in Section 4.4 our view for important future directions in the area of wireless network games.

The content in Sections 4.1– 4.2 relies mostly on [64, 65, 66, 67]. We do not provide herein complete proofs for some of the analytical results. Those could be found in the above references.

4.1 NONCOOPERATIVE TRANSMISSION SCHEDULING IN COLLISION CHANNELS

In this section, we consider a shared uplink in the form of a collision channel. A basic assumption of our user model is that each user has some throughput requirement, which it wishes to sustain with a minimal power investment. The required throughput of each user may be dictated by its application (such as video or voice which may require fixed bandwidth), or mandated by the system. A distinctive feature of our model is that the channel quality between each user and the base station is stochastically varying. For example, the channel quality may evolve as a block fading process [25] with a general underlying state distribution (such as Rayleigh, Rice, and Nakagami-m, see [25]). A user may base its transmission decision upon available indications on the channel state, known as channel state information (CSI). This decision is selfishly made by the individual without any coordination with other users, giving rise to a non-cooperative game. Our focus in this section is on *stationary* transmission strategies, in which the decision whether to transmit or not can depend (only) on the current CSI signal. Non-stationary strategies are naturally harder to analyze, and moreover, their advantage over stationary strategies is not clear in large, distributed and selfish environments.

The technological relevance for our work lies, for example, in Wireless Local Area Network (WLAN) systems where underlying network users have diverse (application-dependent) throughput requirements. The leading standard, namely the 802.11x [1], employs a random access protocol, whose principles are based on the original Aloha. Interestingly, recent IEEE standardization activity (the 802.11n standard) focuses on the incorporation of CSI for better network utilization. This last fact further motivates to study the use of CSI in distributed, self-optimizing user environments.

The main contributions of this section can be summarized as follows:

- Our equilibrium analysis reveals that when the throughput demands are within the network capacity, there exist exactly two Nash equilibrium points in the resulting game.

- One equilibrium is strictly better than the other in terms of power investment for all users, and, in fact, coincides with the socially optimal solution. The performance gap (in terms of the total power investment) between the equilibrium points is potentially unbounded.

- We describe a fully distributed mechanism which converges to the better equilibrium point. The suggested mechanism is natural in the sense that it relies on the user's best response to given network conditions.

The structure of the section is as follows. We first present the general model (Section 4.1.1), and identify basic properties related to stationary transmission strategies. A detailed equilibrium analysis is provided in Section 4.1.2. Section 4.1.3 focuses on the achievable network capacity. In Section 4.1.4, we present a mechanism which converges to the better equilibrium. We discuss several aspects of our results in Section 4.1.5 and highlight further research directions.

4.1.1   THE MODEL AND PRELIMINARIES

We consider a wireless network, shared by a finite set of mobile users $\mathcal{I} = \{1, \ldots, n\}$ who transmit at a fixed power level to a common base station over a shared collision channel. Time is slotted, so that each transmission attempt takes place within slot boundaries that are common to all. A transmission can be successful only if no other user attempts transmission simultaneously. Thus, at each time slot, at most one user can successfully transmit to the base station. To further specify our model, we start with a description of the channel between each user and the base station (Section 4.1.1.1), ignoring the possibility of collisions. In Section 4.1.1.2, we formalize the user objective and formulate the non-cooperative game which arises in a multi-user shared network.

4.1.1.1   The Single-User Channel

Our model for the channel between each user and the base station is characterized by two basic quantities.

a. Channel state information. At the beginning of each time slot k, every user i obtains a channel state information (CSI) signal $\zeta_{i,k} \in \mathcal{Z}_i \subset \mathbb{R}^+$, which provides an indication (possibly partial) of the quality of the current channel between the user and the base station (a larger number corresponds to a better channel quality). We assume that each set $\mathcal{Z}_i$ of possible CSI signals for user i is finite[1] and denote its elements by $\{z_i^1, z_i^2, \ldots, z_i^{x_i}\}$, with $z_i^1 < z_i^2 < \cdots < z_i^{x_i}$.

b. Expected data rate. We denote by $R_i(z_i) > 0$ the expected data rate (in bits per slot) that user i can sustain at any given slot as a function of the current CSI signal $z_i \in \mathcal{Z}_i$. We assume that the function $R_i(z_i)$ strictly increases in z_i.

Throughout this chapter, we make the following assumption:

Assumption 1 *(i) $Z_i = \{\zeta_{i,k}\}_{k=1}^{\infty}$ is an ergodic Markov chain; We denote by π_i the row vector of steady state probabilities of the Markov chain Z_i, and by $\pi_i^m > 0$ its m-th entry corresponding to state $z_i^m \in \mathcal{Z}_i$ (signals with zero steady-state probability are excluded from the set $\mathcal{Z}_i$). (ii) The Markov chains Z_i, $i = 1, \ldots n$, are independent.*

Interpretation: The above model may be used to capture the following network scenario. The quality (or state) of the channel between user i and the base station may vary over time. Let w_i denote an actual channel state for user i at the beginning of some slot (time indexes are omitted here for simplicity). Instead of the exact channel state, user i observes a CSI signal z_i, which is some (possibly noisy) function of w_i. As already noted, larger z_i's indicate better channel conditions. After observing the CSI at the beginning of a slot, user i may respond by adjusting its coding scheme in order to maximize its data throughput on that slot. The expected data rate $R_i(z_i)$ thus takes into account the actual channel state distribution (conditioned on z_i), including possible variation within the slot duration, as well as the coding scheme used by the user. Specifically, let $\tilde{R}_i(w_i, z_i)$ be the expected data rate for channel state w_i, as determined by the coding scheme that corresponds to z_i. Then the expected data rate is given by $R_i(z_i) = \mathbb{E}(\tilde{R}_i(w_i, z_i)|z_i) = \int \tilde{P}_i(w_i|z_i)\tilde{R}_i(w_i, z_i)dw_i$, where $\mathbb{E}$ is the expectation operator and $\tilde{P}_i(w_i|z_i)$ is the conditional probability that the actual channel state is w_i when z_i is observed. Assuming that the actual channel quality forms a Markov chain across slots, this property is clearly inherited by the CSI sequence Z_i.

Our modeling assumptions accommodate, in particular, the so-called block-fading model, which is broadly studied in the literature (see [25, 84] and references therein). Note, however, that our model does not require the i.i.d. assumption, nor does it require the actual channel state to be fixed within each interval.

Example 4.1   Gaussian Channel   Consider the case where the channel quality between user i and the base station evolves as a discrete Markov process, with white Gaussian noise being added to the transmitted signal. Specifically, at each time t, the received signal $y_i(t)$ is given by $y_i(t) = \sqrt{w_i(t)}x_i(t) + n(t)$, where $x_i(t)$ and $w_i(t)$ are, respectively, the transmitted signal and channel gain (which is the physical interpretation for channel quality). Let T be the length of a slot. Then $w_i(t)$

---

[1]This is assumed for convenience only. Note that the channel quality may still take continuous value, which the user reasonably classifies into a finite number of information states.

remains constant within slot boundaries, i.e., $w_i(t) \equiv w_{i,k}, t \in [(k-1)T, kT), k \geq 1$. Suppose user i is able to obtain the underlying channel quality with high precision, namely $\zeta_{i,k} \approx w_{i,k}$. Let S_i be the maximal energy per slot, and let $N_0/2$ be the noise power spectral density. Then if the user can optimize its coding scheme for rates approaching the Shannon capacity, the expected data rate which can be reliably transmitted is given by the well known formula $R_i(z_i) = B_i \log(1 + \frac{S_i}{N_0} z_i)$, where B_i is the bandwidth.

4.1.1.2 User Objective and Game Formulation

We turn now to describe the user objective and the non-cooperative game which arises as a consequence of the user interaction over the collision channel.

Basic Definitions. We associate with each user i has a throughput demand ρ_i (in bits per slot) which it[2] wishes to deliver over the network. The objective of each user is to minimize its average transmission power (which is equivalent in our model to the average rate of transmission attempts, as users transmit at a fixed power level), while maintaining the effective data rate at (or above) this user's throughput demand. We further assume that users always have packets to send, yet they may postpone transmission to a later slot to accommodate their required throughput with minimal power investment.

A general transmission schedule, or strategy, σ_i for user i specifies a transmission decision at each time instant, based on the available information that includes the CSI signals and (possibly) the transmission history for that user. A transmission decision may include randomization (i.e., transmit with some positive probability). Since this section focuses on stationary transmission strategies, we will not bother with a formal definition of a general strategy. For our purpose, it suffices to assume that the collection of user strategies $(\sigma_i)_{i \in \mathcal{I}}$ together with the channel description, induce a well defined stochastic process of user transmissions.

Obviously, each user's strategy σ_i directly affects other users' performance through the commonly shared medium. The basic assumption of our model is that users are self-optimizing and are free to determine their own transmission schedule in order to fulfill their objective. We further assume that users are unable to coordinate their respective decisions. This situation is modeled and analyzed in this section as a non-cooperative game between the n users.

We denote by $\sigma = (\sigma_1, \ldots, \sigma_n)$ the strategy-profile comprised of all users' strategies. The notation σ_{-i} is used for the transmission strategies of all users but the i-th one. For each user i, let $p_i(\sigma)$ be the average transmission rate (or transmission probability), and let $r_i(\sigma)$ be the expected average throughput as determined by the user's own strategy σ_i and by the strategies of all other users σ_{-i}. Further denote by $c_{i,k}$ the indicator random variable which equals one if user i transmits at slot k and zero, otherwise, and by $r_{i,k}$ the number of data bits *successfully* transmitted by user i at

---

[2]The *user* here should be interpreted as the algorithm that manages the transmission schedules and is accordingly referred to in the third person neuter.

the same slot. Then

$$p_i(\sigma) = \lim_{K \to \infty} \mathbb{E}^\sigma \Big(\frac{1}{K} \sum_{k=1}^{K} c_{i,k}\Big), \tag{4.2}$$

$$r_i(\sigma) = \lim_{K \to \infty} \mathbb{E}^\sigma \Big(\frac{1}{K} \sum_{k=1}^{K} r_{i,k}\Big), \tag{4.3}$$

where $\mathbb{E}^\sigma$ stands for the expectation operator under the strategy-profile σ. If the limit in (4.2) does not exist, we may take the lim sup instead, and similarly the lim inf in (4.3).

A Nash equilibrium (NE) is a strategy-profile $\sigma = (\sigma_1, \ldots, \sigma_n)$, which is self-sustaining in the sense that all throughput constraints are met, and no user can lower its transmission rate by unilaterally modifying its transmission strategy. Formally,

Definition 4.2 Nash equilibrium point   A strategy-profile $\sigma = (\sigma_1, \ldots, \sigma_n)$ is a *Nash equilibrium point* if

$$\sigma_i \in \underset{\tilde{\sigma}_i}{\operatorname{argmin}} \{p_i(\tilde{\sigma}_i, \sigma_{-i}) : r_i(\tilde{\sigma}_i, \sigma_{-i}) \geq \rho_i\}. \tag{4.4}$$

The transmission rate p_i can be regarded as the cost which the user wishes to minimize. Using game-theoretic terminology, a Nash equilibrium is a strategy-profile $\sigma = (\sigma_1, \ldots, \sigma_n)$ so that each σ_i is a *best response* of user i to σ_{-i}, in the sense that the user's cost is minimized.

Our focus in this section is on locally *stationary* transmission strategies, in which the user's decision whether to transmit or not can depend (only) on its current CSI signal (for simplicity, we shall henceforth refer to such strategy just as stationary strategy, yet recall that local information only is employed by each user). A formal definition for a stationary strategy is provided below.

Definition 4.3 Stationary strategies   A *stationary strategy* for user i is a mapping $\sigma_i : \mathcal{Z}_i \to [0, 1]$. Equivalently, a stationary strategy will be represented by an x_i-dimensional vector $\mathbf{s}_i = (s_i^1, \ldots, s_i^{x_i}) \in [0, 1]^{x_i}$, where the m-th entry corresponds to the user i's transmission probability when the observed CSI signal is z_i^m. For example, the vector $(0, \ldots, 0, 1)$ represents the strategy of transmitting (w.p. 1) only when the CSI signal is the highest possible. Note that the transmission probability in a slot, which is a function of $\mathbf{s}_i$ only, is given by

$$p_i(\mathbf{s}_i) = \sum_{m=1}^{x_i} s_i^m \pi_i^m. \tag{4.5}$$

Let $\mathbf{s} \overset{\triangle}{=} (\mathbf{s}_1, \ldots, \mathbf{s}_n)$ denote a stationary strategy-profile for all users. Evidently, the probability that no user from the set $\mathcal{I} \backslash i$ transmits in a given slot is given by $\prod_{j \neq i}(1 - p_j(\mathbf{s}_j))$. Since the transmission decision of each user is independent of the decisions of other users, the expected average rate $r_i(\mathbf{s}_i, \mathbf{s}_{-i})$ is given by

$$r_i(\mathbf{s}_i, \mathbf{s}_{-i}) = \left[\sum_{m=1}^{x_i} s_i^m \pi_i^m R_i(z_i^m) \right] \prod_{j \neq i} (1 - p_j(\mathbf{s}_j)), \tag{4.6}$$

where the expression $\sum_{m=1}^{x_i} s_i^m \pi_i^m R_i(z_i^m)$ stands for the average rate which is obtained in a collision-free environment under the same strategy $\mathbf{s}_i$.

Remark 4.4   As noted above, we restrict attention here to stationary strategies. When the CSI process is i.i.d., it may be shown that a Nash equilibrium in stationary strategies is, in fact, a Nash equilibrium in general strategies. For more general state processes (e.g., Markovian), this need not be the case, and the restriction to stationary strategies is upheld for simplicity of implementation.

Threshold Strategies. A subclass of stationary strategies which is central in our analysis is defined below.

Definition 4.5 Threshold strategies   A *threshold strategy* is a stationary strategy of the form $\mathbf{s}_i = (0, 0, \ldots, 0, s_i^{m_i}, 1, 1 \ldots, 1)$, $s_i^{m_i} \in (0, 1]$, where $z_i^{m_i}$ is a threshold CSI level above which user i always transmits, and below which it never transmits. An important observation, which we

summarize next, is that users should always prefer threshold strategies.

Lemma 4.6   *Assume that all users access the channel using a stationary strategy. Then a best response strategy of any user i is always a threshold strategy.*

As a result of the above lemma, we may analyze the non-cooperative game by restricting the strategies of each user i to the set of threshold strategies, denoted by T_i. We proceed by noting that every threshold strategy can be identified with a unique *scalar* value $p_i \in [0, 1]$, which is the transmission probability in every slot, i.e., $p_i \equiv p_i(\mathbf{s}_i)$. More precisely:

Lemma 4.7   *The mapping $\mathbf{s}_i = (0, 0, \ldots, 0, s_i^{m_i}, 1, 1 \ldots, 1) \in T_i \to p_i \equiv p_i(\mathbf{s}_i) \in [0, 1]$, is a surjective (one-to-one and onto) mapping from the set of threshold strategies T_i to the interval $[0, 1]$.*

Proof. The claim follows directly from (4.5), upon recalling that $\sum_m \pi_i^m = 1$ and $\pi_i^m > 0$ by assumption. Note first that under a threshold strategy $\mathbf{s}_i$, $p_i(\mathbf{s}_i) = s_i^{m_i} \pi_i^{m_i} + \sum_{m=m_i+1}^{x_i} \pi_i^m$. Evidently, $0 \leq p_i(\mathbf{s}_i) \leq \sum_m \pi_i^m = 1$. Conversely, every $p_i \in [0, 1]$ corresponds to a unique threshold strategy as follows: Given p_i, the corresponding m_i is such that $\sum_{m=m_i+1}^{x_i} \pi_i^m < p_i$ and $\sum_{m=m_i}^{x_i} \pi_i^m \geq p_i$; the transmission probability for the threshold CSI is given by $s_i^{m_i} = \frac{p_i - \sum_{m=m_i+1}^{x_i} \pi_i^m}{\pi_i^{m_i}}$. $\qquad\square$

Given this mapping, the stationary policy of each user will be henceforth represented through a scalar $p_i \in [0, 1]$, which uniquely determines the CSI threshold and its associated transmission probability, denoted by $z_i^{m_i}(p_i)$ and $s_i^{m_i}(p_i)$, respectively. Consequently, the user's expected throughput per slot in a collision free environment, denoted by H_i, can be represented as a function of p_i only, namely

$$H_i(p_i) \stackrel{\triangle}{=} s_i^{m_i}(p_i)\pi_i^{m_i}(p_i)R_i(z_i^{m_i}(p_i)) + \sum_{m=m_i(p_i)+1}^{x_i} \pi_i^m R_i(z_i^m), \tag{4.7}$$

where $m_i(p_i)$ denotes the index of the threshold CSI and $\pi_i^{m_i}(p_i)$ denotes its probability. This function will be referred to as the *collision-free rate function*. Using this function, we may obtain an explicit expression for the user's average throughput, as a function of $\mathbf{p} = (p_1, \ldots, p_n)$, namely

$$r_i(p_i, \mathbf{p}_{-i}) = H_i(p_i) \prod_{j \neq i}(1 - p_j). \tag{4.8}$$

Example 4.8   Null CSI   A special important case is when no CSI is available. This corresponds to $x_i = 1$ in our model. In this case, the collision-free rate function is simply $H_i(p_i) = \bar{R}_i p_i$, where $\bar{R}_i = R_i(z_i^1)$ is the expected data rate that can be obtained in any given slot.

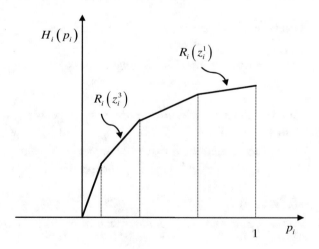

Figure 4.2: An example of the collision-free rate function $H_i(p_i)$. In this example, there are four CSI signals. Note that the slope of $H_i(p_i)$ is exactly the rate of the threshold CSI which corresponds to p_i.

Some useful properties of the rate function (4.7) are summarized in the next lemma.

Lemma 4.9   *The collision-free rate function H_i satisfies the following properties.*

(i) $H_i(0) = 0$.

(ii) $H_i(p_i)$ *is a continuous and strictly increasing function over* $p_i \in [0, 1]$.

(iii) $H_i(p_i)$ *is concave.*

Proof. Noting (4.5), $p_i = 0$ means no transmission at all, thus an average rate of zero. It can be easily seen that $H_i(p_i)$ in (4.7) is a piecewise-linear (thus continuous), strictly increasing function. As to concavity, note that the slope of H_i is determined by $R_i(z_i^{m_i})$ which decreases with p_i (see Figure 4.2), as m_i decreases in p_i (from Eq. (4.7)) and z_i^m is increasing in m (by definition).   □

4.1.2   EQUILIBRIUM ANALYSIS

In this section, we describe several basic properties of the Nash equilibria under stationary transmission strategies. In Section 4.1.2.1, we define the equilibrium equations, which must be satisfied at every feasible equilibrium point and are essential for our analysis. In Section 4.1.2.2, we characterize the feasible region of throughput demands, which is the set of vectors $\rho = (\rho_1, \ldots, \rho_n)$ for which a Nash equilibrium point exists. We further show that there exist exactly two equilibrium point for each ρ within the feasible region. We then prove that one equilibrium is better than the other for all users in terms of the invested power (Section 4.1.2.3) and quantify the efficiency loss incurred by selfish behavior of users (Section 4.1.2.4). We conclude this section by commenting on computational aspects related to the efficient calculation of equilibria.

4.1.2.1   The Equilibrium Equations

A key property which is useful for the analysis is that every Nash equilibrium point can be represented via a set of n equations in the n variables $\mathbf{p} = (p_1, \ldots, p_n)$. This is summarized in the next proposition.

Proposition 4.10   The equilibrium equations   *A strategy-profile* $\mathbf{p} = (p_1, \ldots, p_n)$ *is a Nash equilibrium point if and only if it solves the following set of equations*

$$r_i(p_i, \mathbf{p}_{-i}) = H_i(p_i) \prod_{j \neq i}(1 - p_j) = \rho_i, \quad i \in \mathcal{I}. \tag{4.9}$$

Proof. Adapting the Nash equilibrium definition (4.4) to stationary threshold strategies, a NE is a strategy-profile $\mathbf{p} = (p_1, \ldots, p_n)$ such that

$$p_i = \min\left\{\tilde{p}_i \in [0, 1], \quad \text{subject to } r_i(\tilde{p}_i, \mathbf{p}_{-i}) \geq \rho_i\right\}, \quad i \in \mathcal{I}, \tag{4.10}$$

where r_i is defined in (4.8). Since $r_i(\tilde{p}_i, \mathbf{p}_{-i})$ is strictly increasing in $\tilde{p}_i$ (by Lemma 4.9), (4.10) is equivalent to $r_i(p_i, \mathbf{p}_{-i}) = \rho_i, i \in \mathcal{I}$, which is just (4.9).   □

Due to the above result, we shall refer to the set of equations (4.9) as the *equilibrium equations.*

4.1.2.2 Two Equilibria or None

Obviously, if the overall throughput demands of the users are too high, there cannot be an equilibrium point since the network naturally has limited traffic capacity (the capacity of the network will be considered in Section 4.1.3).

Denote by $\rho = (\rho_1, \dots, \rho_n)$ the vector of throughput demands, and let Ω be the set of feasible vectors ρ, for which there exists at least one Nash equilibrium point (equivalently, for which there exists a feasible solution to (4.9)). Figure 4.3 illustrates the set of feasible throughput demands for a simple two-user case, with $H_i(p_i) = p_i$.

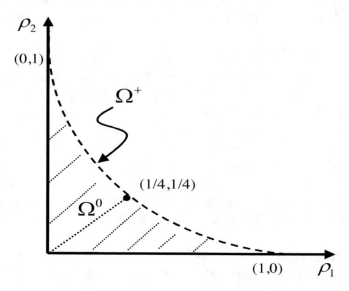

Figure 4.3: The set of feasible throughput demands for a two user network with $H_i(p_i) = p_i, i = 1, 2$.

To specify some structural properties of Ω, it is convenient to define the set of basis vectors $\hat{\Omega}$, where each $\hat{\rho} = (\hat{\rho}_1, \dots, \hat{\rho}_n) \in \hat{\Omega}$ is such that $\hat{\rho}_i > 0$ for every $i \in \mathcal{I}$ and $\|\hat{\rho}\|_2 = 1$, i.e., $\sum_i \hat{\rho}_i^2 = 1$.

Proposition 4.11 *The feasible set Ω obeys the following properties.*
(i) Closed cone structure: For every $\hat{\rho} \in \hat{\Omega}$, $\alpha\hat{\rho} \in \Omega$ for all $\alpha \in [0, \alpha(\hat{\rho})]$.
(ii) Let $\rho \le \tilde{\rho}$ be two throughput demand vectors. Then if $\tilde{\rho} \in \Omega$, it follows that $\rho \in \Omega$.

In particular, note that Ω is a closed set with nonempty interior. We can now specify the number of equilibrium points for any throughput demand vector $\rho = (\rho_1, \dots, \rho_n)$. When throughput demands are within the feasible region Ω, we establish that there are exactly *two* Nash equilibria.

Theorem 4.12 *Consider the non-cooperative game model under stationary transmission strategies. Let Ω be the set of feasible throughput demand vectors $\rho = (\rho_1, \dots, \rho_n)$, and let Ω^0 be its (non-empty) interior. Then for each $\rho \in \Omega^0$, there exist exactly two Nash equilibria.*

The idea of the proof is to reduce the equation set (4.9) to a single scalar equation in a single variable p_i, for some arbitrarily chosen user i. The right-hand side of the equation remains ρ_i; unimodality of the left-hand side will establish the required result; further details can be found in [64, 66].

4.1.2.3 The Energy Efficient Equilibrium

Going beyond the basic questions of existence and number of equilibrium points, we wish to further characterize the properties of the equilibrium points. In particular, we are interested here in the following question: How do the two equilibrium points compare: is one "better" than the other? The next theorem shows that, indeed, one equilibrium point is power-superior for all users.

Theorem 4.13   *Assume that the throughput demand vector ρ is within the feasible region Ω^0, so that there exist two equilibria in stationary strategies. Let $\mathbf{p}$ and $\tilde{\mathbf{p}}$ be these two equilibrium points. If $p_i < \tilde{p}_i$ for some user i, then $p_j < \tilde{p}_j$ for every $j \in \mathcal{I}$.*

Proof. Define $a_{ik} \overset{\triangle}{=} \frac{\rho_i}{\rho_k}$. For every user $k \neq i$ divide the i-th equation in the set (4.9) by the k-th one. We obtain

$$a_{ik} = \frac{H_i(p_i)(1 - p_k)}{H_k(p_k)(1 - p_i)} < \frac{H_i(\tilde{p}_i)(1 - p_k)}{H_k(p_k)(1 - \tilde{p}_i)}, \tag{4.11}$$

since H_i is increasing. Now since $\frac{H_i(\tilde{p}_i)(1-\tilde{p}_k)}{H_k(\tilde{p}_k)(1-\tilde{p}_i)} = a_{ik}$, it follows that $\frac{(1-\tilde{p}_k)}{H_k(\tilde{p}_k)} < \frac{(1-p_k)}{H_k(p_k)}$. Since H_k is increasing in p_k, we conclude from the last inequality that $p_k < \tilde{p}_k$.   □

The last result is significant from the network point of view. It motivates the design of a network mechanism that will avoid the inferior equilibrium point, which is wasteful for *all* users. This will be our main concern in Section 4.1.4. Henceforth, we identify the better equilibrium point as the *Energy Efficient Equilibrium (EEE)*.

We now turn to examine the quality of the EEE relative to an appropriate social cost. Recall that each user's objective is to minimize its average transmission rate subject to a throughput demand. Thus, a natural performance criterion for evaluating any strategy-profile $\mathbf{s} = (\mathbf{s}_1, \ldots, \mathbf{s}_n)$ (in particular, an equilibrium strategy-profile) is given by the sum of the user's average transmission rates induced by $\mathbf{s}$, namely

$$Q(\mathbf{s}) = \sum_i p_i(\mathbf{s}_i). \tag{4.12}$$

The next theorem addresses the quality of the EEE with respect to that criterion.

Theorem 4.14   *Let $\mathbf{p}$ be an EEE. Then $\sum_i p_i \leq 1$.*

An immediate conclusion from the above theorem is that the overall power investment at the EEE is bounded, as the sum of transmission probabilities is bounded. This means, in particular, that the average transmission power of all users is bounded by the maximal transmission power of a single station.

4.1.2.4 Social Optimality and Efficiency Loss

We proceed to examine the extent to which selfish behavior affects system performance. That it, we are interested to compare the quality of the obtained equilibrium points to the centralized, system-optimal solution (still restricted to stationary strategies). To that end, we shall use the notions of *price of anarchy* (PoA) (see Chapter 3, Section 3.1.3) and *price of stability* (PoS). Recall that the *price of anarchy* (PoA) is (an upper bound on) the performance ratio (in terms of a relevant social performance measure) between the global optimum and the *worst* Nash equilibrium, while the *price of stability* (PoS) is (an upper bound on) the performance ratio between the global optimum and the *best* Nash equilibrium.

Returning to our specific network scenario, consider the case where a central authority, which is equipped with user characteristics $\mathbf{H} = (H_1, \ldots, H_n)$ and $\rho = (\rho_1, \ldots, \rho_n)$ can enforce a stationary transmission strategy for every user $i \in \mathcal{I}$. We consider (4.12) as the system-wide performance criterion, and we compare the performance of this optimal solution to the performance at the Nash equilibria. A socially optimal strategy-profile denoted $\mathbf{s}^*(\mathbf{H}, \rho)$, is a strategy that minimizes (4.12), while obeying all user throughput demands ρ_i. Similarly, denote by $\mathbf{s}^{\mathbf{a}}(\mathbf{H}, \rho)$ and $\mathbf{s}^{\mathbf{b}}(\mathbf{H}, \rho)$ the multi-strategies at the better NE and at the worse NE, respectively. Then the PoA and PoS are given by

$$PoA = \sup_{\mathbf{H}, \rho} \frac{Q(\mathbf{s}^{\mathbf{b}}(\mathbf{H}, \rho))}{Q(\mathbf{s}^*(\mathbf{H}, \rho))}, \quad PoS = \sup_{\mathbf{H}, \rho} \frac{Q(\mathbf{s}^{\mathbf{a}}(\mathbf{H}, \rho))}{Q(\mathbf{s}^*(\mathbf{H}, \rho))}. \tag{4.13}$$

We next show that the PoA is generally unbounded, while the PoS is always one.

Theorem 4.15   *Consider the non-cooperative game under stationary transmission strategies. Then (i) the PoS is always one, and (ii) the PoA is generally unbounded.*

Proof. (i) This claim follows immediately, noting that (1) the socially optimal stationary strategy is a threshold strategy (by applying a similar argument to the one used in Lemma 4.6), and (2) the socially optimal stationary strategy obeys the equilibrium equations (4.9) (following a similar argument to the one used in Proposition 4.10). Hence, by Proposition 4.10, the optimal solution is also an equilibrium point. Equivalently, this means that $PoS = 1$.

(ii) We establish that the price of anarchy is unbounded by means of an example. Consider a network with n identical users with $H_i(p_i) = \bar{R}p_i$ (this collision-free rate function corresponds to users who cannot obtain any CSI). Each user's throughput demand is $\rho_i = \epsilon \to 0$. Recall that the throughput demands are met with equality at every equilibrium point (Proposition 4.10). Then, by symmetry, we obtain a single equilibrium equation, namely $\bar{R}p(1 - p)^{n-1} = \epsilon$. As ϵ goes to zero, the two equilibria are $p_a \to 1$ and $p_b \to 0$. Obviously, the latter point is also a social optimum; it is readily seen that the price of anarchy here equals in the limit to $\frac{p_a}{p_b} \to \infty$.   □

The above theorem clearly motivates the need for a mechanism that will induce the EEE, as this equilibrium point coincides with the socially-optimal solution, while the gap between the two equilibria could be arbitrarily large. Such mechanism is considered in Section 4.1.4.

4.1.3   ACHIEVABLE CHANNEL CAPACITY

The aim of this section is to provide explicit lower bounds for the achievable channel capacity. The term "capacity" is used here for the total throughput (normalized to successful transmission per slot), which can be obtained in the network. We focus here on the case where users have no CSI, and then relate our result to general CSI.

Consider the null-CSI model, where no user can observe any CSI (see the example at the end of Section 4.1.1). Recall that the collision-free rate in this case is given by $H_i(p_i) = \bar{R}_i p_i$, where $\bar{R}_i$ is the expected data rate in case of a successful transmission. Define $y_i \triangleq \frac{\rho_i}{\bar{R}_i}$, which we identify henceforth as the *normalized throughput* demand for user i; indeed, y_i stands for the required rate of successful transmissions. Then the equilibrium equations (4.9) become

$$p_i \prod_{j \neq i}(1 - p_j) = y_i, \quad 1 \leq i \leq n. \tag{4.14}$$

We shall first consider the symmetric case, i.e., $y_i = y$ for every user i, and then relate the results to the general non-symmetric case. The theorem below establishes the conditions for the existence of an equilibrium point in the symmetric null-CSI case.

Theorem 4.16   Symmetric users   *Let $y_i = y$ for every $1 \leq i \leq n$. Then (i) A Nash equilibrium exists if and only if*

$$ny \leq (1 - \frac{1}{n})^{n-1}. \tag{4.15}$$

(ii) In particular, a Nash equilibrium exists if $ny \leq e^{-1}$.

Proof. (i) By dividing the equilibrium equations (4.14) of any two users, it can be seen that every symmetric-users equilibrium satisfies $p_i = p_j = p$ ($\forall i, j$). Thus, the equilibrium equations (4.14) reduce to a single (scalar) equation:

$$h(p) \triangleq p(1 - p)^{n-1} = y. \tag{4.16}$$

We next investigate the function $h(p)$. Its derivative is given as $h'(p) = (1 - p)^{n-2}(1 - np)$. It can be seen that the maximum value of the function $h(p)$ is obtained at $p = 1/n$. An equilibrium exists if and only if the maximal value of $h(p)$ is greater than y. Substituting the maximizer $p = 1/n$ in (4.16) implies the required result.

(ii) It may be easily verified that the right-hand side of (4.15) decreases with n. Since $\lim_{n \to \infty}(1 - \frac{1}{n})^{n-1} = e^{-1}$, the claim follows from (i).   □

In [66], it is shown that the simple bound obtained above holds for non-symmetric users as well, implying that the symmetric case is the worst in terms feasible channel utilization. Formally,

Theorem 4.17   Asymmetric users   *For any set of n null-CSI users with normalized throughput demands* $\{y_i\}$, *an equilibrium point exists if*

$$\sum_{i=1}^{n} y_i \leq (1 - \frac{1}{n})^{n-1}. \tag{4.17}$$

The quantity e^{-1} is also the well-known maximal throughput of a slotted Aloha system with Poisson arrivals and an infinite set of nodes [19]. In our context, if the normalized throughput demands do not exceed e^{-1}, an equilibrium point is guaranteed to exist. Thus, in a sense, we may conclude that noncooperation of users, as well as restricting users to stationary strategies, do not reduce the capacity of the collision channel.

We conclude this section by noting that Eq. (4.17) serves as a global sufficient condition for the existence of an equilibrium point, which holds for any level of channel observability. This observation follows by showing that the capacity can only increase when users obtain channel state information (see [66] for details).

4.1.4   BEST-RESPONSE DYNAMICS

A Nash equilibrium point for our system represents a strategically stable working point, from which no user has incentive to deviate unilaterally. Still, the question of if and how the system arrives at an equilibrium remains open. Furthermore, since our system has two Nash equilibria with one (the EEE) strictly better than the other, it is of major importance (from the system viewpoint, as well as for each individual user) to employ mechanisms that converge to the better equilibrium rather than the worse.

The distributed mechanism we consider here relies on a user's best-response (BR), which is generally the optimal user reaction to a given network condition (see [46]). Specifically, the best response of a given user is a transmission probability which brings the obtained throughput of that user (given other user strategies) to its throughput demand ρ_i. Accordingly, observing (4.8), the best response of user i for any strategy-profile $\mathbf{p} = (p_1, \ldots, p_n)$ is given by

$$p_i := H_i^{-1} \left(\frac{\rho_i}{\prod_{j \neq i}(1 - p_j)} \right), \tag{4.18}$$

where H_i^{-1} is the inverse function of the collision-free rate function H_i (if the argument of H_i^{-1} is larger than maximal value of H_i, p_i can be chosen at random). Note that H_i^{-1} is well defined, since H_i is continuous and monotone (Lemma 4.9). It is important to notice that each user is not required to be aware of the transmission probability of every other user. Indeed, only the overall idle probability of other users $\prod_{j \neq i}(1 - p_j)$ is required in (4.18). Our mechanism can be described as follows. *Each user updates its transmission probability from time to time through its best response (4.18). The update times of each user need not be coordinated with other users.*

This mechanism reflects what greedy, self-interested users would naturally do: repeatedly observe the current network situation and react to bring their costs to a minimum.

The analysis of the best-response mechanism will be carried out under the following assumptions that apply throughout this section.

Assumption 2

(i) Fixed Demands: *The user population and the users' throughput requirements* $\rho_1, \ldots, \rho_n$ *are fixed. Furthermore,* $\rho_1, \ldots, \rho_n$ *are within the feasible throughput region.*

(ii) Persistent Updates: *Each user updates its transmission probabilities using Eq. (4.18) at arbitrarily chosen time instants, and the number of updates is unbounded.*

Furthermore, in order to guarantee convergence to the EEE (denoted henceforth as $\mathbf{p}^a$), we must impose restrictions on the initial strategy profile of the users, denoted $\mathbf{p}^0$. Such restrictions are clearly essential due to the inherent non-uniqueness of the equilibrium in our model; indeed, if we start at the *worse* equilibrium point (denoted henceforth as $\mathbf{p}^b$), we will stay there indefinitely under BR. For initial conditions above $\mathbf{p}^b$, we might observe throughput collapse, as some users increase their transmission probabilities to their maximal values of 1.

To specify the required condition, let us define the following sets of strategy profiles:

$$\Pi_1 = \big\{ \mathbf{p} \in [0, 1]^n : r_i(\mathbf{p}) \geq \rho_i \text{ for every } i, \text{ and}$$
$$p_j < p_j^b \text{ for some } j \big\};$$
$$\Pi_0 = \big\{ \mathbf{p} \in [0, 1]^n : \mathbf{p} \leq \bar{\mathbf{p}} \text{ for some } \bar{\mathbf{p}} \in \Pi_1 \big\}.$$

Our main convergence result is summarized below.

Theorem 4.18 BR Convergence   *Let the initial request probabilities* $\mathbf{p}^0 = (p_1^0 \ldots, p_n^0)$ *satisfy* $\mathbf{p}^0 \in \Pi_0$. *Then the best response dynamics asymptotically converges to the* better *equilibrium point* $\mathbf{p}^a$. The

proof proceeds by applying a "sandwich" argument. We first show that BR converges monotonously (from below) to $\mathbf{p}^a$ when started at $\mathbf{p}^0 = 0$. We then show that it converges monotonously (from above) to $\mathbf{p}^a$ when started with $\mathbf{p}^0$ in Π_1. Finally, we conclude by monotonicity of the BR that convergence must occur all for initial conditions between 0 and Π_1, namely for all $\mathbf{p}^0 \in \Pi_0$. Details can be found in [67].

We briefly list here some considerations regarding the presented mechanism.

1. It is important to notice that each user is not required to be aware of the transmission probability of every other user. Indeed, only the overall idle probability of other users $\prod_{j \neq i}(1 - p_j)$ is required in (4.18). This quantity could be estimated by each user by monitoring the channel utilization.

2. The update-rule (4.18) entails the notion of a quasi-static system, in which each user responses to the steady state reached after preceding user updates. This assumption approximates a natural scenario where users update their transmission probabilities at much slower time-scales than their respective transmission rates. A recent paper [92] demonstrates that the quasi-static assumption can be relaxed while still ensuring convergence of user dynamics, by employing stochastic approximation tools.

In [66], we show that the appealing convergence properties continue to hold when the user population changes over time. We notice that the convergence results obtained in the section would still hold for a relaxed variation of (4.18), given by

$$p_i := \beta_i H_i^{-1} \left(\frac{\rho_i}{\prod_{j \neq i}(1 - p_j)} \right) + (1 - \beta_i)p_i, \tag{4.19}$$

where $0 \leq \beta_i \leq 1$. This update rule can be more robust against inaccuracies in the estimation of $\prod_{j \neq i}(1 - p_j)$, perhaps at the expense of slower convergence to the desired equilibrium.

Our convergence results are obviously idealized and should be supplemented with further analysis of the effect of possible deviations from the model and possible remedies. In case that a worst equilibrium point is reached (or no equilibrium is obtained after a reasonably long time), users can reset their probabilities and restart the mechanism (4.18) for converging to the better equilibrium. This procedure resembles the basic ideas behind TCP protocols. The exact schemes for detecting operation at suboptimal equilibria, and consequently directing the network to the EEE, are beyond the scope of the present research.

4.1.5   DISCUSSION

We briefly discuss here some consequences of our results, emphasizing network management aspects, and highlight some interesting future research directions. Our equilibrium analysis has revealed that within the feasible region the system has two Nash equilibrium points with one strictly better than the other. The better equilibrium (the EEE) is socially optimal, hence the network should ensure that users indeed operate at that equilibrium. An important step in this direction is the above suggested distributed mechanism which converges to the EEE. It should be mentioned, however, that fluctuations in the actual system might clearly bring the network to an undesired equilibrium. Hence, centralized management (based on user feedbacks) may still be required to identify the possible occurrence of the worse equilibrium, and then direct the network to the EEE. Possible mechanisms for this purpose remain a research direction for the future.

In this section, we mainly considered the throughput demands ρ_i as determined by the user itself. Alternatively, ρ_i may be interpreted as a bound on the allowed throughput which is imposed by the network (as part of a resource allocation procedure). The advantage of operating in this "allocated-rate" mode is twofold. First, the network can ensure that user demands do not exceed the network capacity (e.g., by restricting the allocated rate, or through call admission control). Second, users can autonomously reach an efficient working point without network involvement, as management overhead is reduced to setting the user rates only. The rate allocation phase (e.g., through service level agreements) is beyond the scope of the present model.

Another important comment relates to elastic users that may lower their throughput demand based on a throughput–power tradeoff. An obvious effect of demand elasticity would be to lower the throughput at inefficient equilibria. It remains to be verified whether other properties established here remain valid in this case.

The framework and results of this section may be extended in several ways. An interesting research extension is to consider non-stationary user strategies. A central question is whether the system benefits from the use of more complex policies by selfish individuals. The incorporation of non-stationary strategies seems to add considerable difficulty to the analysis and may require more elaborate game theoretic tools than the ones used here.

4.2 NONCOOPERATIVE POWER CONTROL IN COLLISION CHANNELS

This section considers the combined power-control and transmission scheduling problem in a time-slotted collision channel. At every time slot, each user may observe its own channel quality and decide whether to transmit or not, and if so at which power level (chosen from a discrete set). Our focus in this section is on local *stationary* transmission strategies, in which this decision can depend (only) on the current channel state of the mobile. As in Section 4.1, we assume that each mobile user has some throughput requirement, which it wishes to sustain with a minimal power investment. Users cannot coordinate their transmissions, and adjust their transmission decisions to minimize their power investment based on network conditions. This situation is modeled and analyzed as a noncooperative game between the mobiles. The basic objective of this section is to obtain structural properties for the associated Nash equilibria and investigate the effects of the physical parameters (e.g., the channel state distribution, the available power levels) on equilibrium performance.

The main findings of this section are the following:

- In contrast to Section 4.1, there are possibly more than two equilibria for the underlying game. Importantly, the equilibria are completely ordered in terms of the per-user power investment; hence, as before there is one equilibrium which is best for all. This equilibrium can be reached through best response dynamics that requires minimal information structure for each mobile.

- On the negative side, the power-efficient equilibrium is usually inferior to the centrally assigned power schedule. In addition, we demonstrate that the freedom given to users in the form of multiple power levels might have negative effects on network performance, in terms of both channel capacity and overall power consumption. These properties could be regarded as Braess-like paradoxes (see Chapter 3, Section 3.1.3) in wireless networks.

The structure of the section is as follows. We first present the general model (Section 4.2.1) and define the Nash equilibrium point of the noncooperative game. Structural results of the Nash equilibria are derived in Section 4.2.2. In addition, we show that if the required rates are feasible, there exists an equilibrium point which is uniformly best for all users in terms of the power investment. Accordingly, we suggest in Section 4.2.3 a simple distributed mechanism that converges to that equilibrium. In Section 4.2.4, we study the efficiency loss incurred by selfish user behavior and identify a couple of Braess-like paradoxes. Section 4.2.5 discusses some consequences of our results, and outlines further research.

4.2.1   THE MODEL

We consider a time-slotted wireless network, shared by a finite set of mobile users $\mathcal{I} = \{1, \ldots, n\}$ who transmit to a common base station over a shared collision channel. A finite set of power levels[3] $\mathcal{Q}_i = \{Q_i^0, Q_i^1, \ldots, Q_i^{J_i}\}$ is available to each mobile i, where $0 = Q_i^0 < Q_i^1, \cdots < Q_i^{J_i}$. A transmission at any (positive) power level is successful only if no other user attempts transmission simultaneously.

As in Section 4.1, we first describe the channel characteristics between each user and the base station (Section 4.2.1.1), ignoring the possibility of collisions. In Section 4.2.1.2, we formalize the user objective and formulate the non-cooperative power control game between the users, and the associated Nash equilibrium.

4.2.1.1   The Single-User Channel

Our model for the channel between each user and the base station is characterized by two basic quantities.

a. Channel State. We use an identical channel state model as in Section 4.1.1.1. For simplicity, we shall often use the term "channel state" when actually referring to the associated CSI signal, yet we recall that partial information only may be available to the user. For completeness, we briefly repeat the basic modeling assumptions used in Section 4.1 with regard to the channel state process.

We assume that the channel state (or quality) between mobile i and the base station evolves as an ergodic Markov chain Z_i, taking values in a finite set $\mathcal{Z}_i = (z_i^1, z_i^2, \ldots, z_i^{x_i})$ of x_i states. For convenience, we shall assume that the states are ordered from worst (z_i^1) to best $(z_i^{x_i})$ and denote this relation by $z_i^1 < z_i^2 < \ldots < z_i^{x_i}$. The Markov chains $Z_i, i = 1 \ldots n$, are assumed to be independent. We denote by π_i the row vector of steady state probabilities of the Markov chain Z_i, and by π_i^m its m-th entry corresponding to state $z_i^m \in \mathcal{Z}_i$.

b. Expected data rate. Let $R_i^{m,j} \geq 0$ denote the expected data rate (in bits per second) that user i can sustain at a given slot as a function of the current channel state z_i^m and the power level Q_i^j assigned for the transmission. We assume that the data rate strictly increases with the channel quality, and

---

[3]A continuum of power levels can be treated as well, and in fact turns out to be analytically simpler. Here we choose to focus on the finite case which is more realistic.

further that it strictly increases with the transmission power. That is,

$$R_i^{1,j} < R_i^{2,j} < \cdots < R_i^{x_i,j}, \quad j \in \{1, \ldots J_i\} \tag{4.20}$$
$$R_i^{m,1} < R_i^{m,2} < \cdots < R_i^{m,J_i}, \quad m \in \{1, \ldots x_i\}; \tag{4.21}$$

naturally, $R_i^{m,0} = 0$, since $Q_i^0 = 0$ by definition.

Example 4.19   Assume that white Gaussian noise with a spectral density of $N_0/2$ is added to the transmitted signal. Then if a user can optimize its coding scheme to approach the Shannon capacity, the average rate that can be sustained is given by the following rate function

$$R_i^{m,j} = \log(1 + z_i^m Q_i^j / N_0). \tag{4.22}$$

We shall consider this specific rate function in Section 4.2.4.

4.2.1.2 User Objective and Game Formulation

In this subsection, we describe the user objective and the non-cooperative game which arises as a consequence of the user interaction over the collision channel. We characterize stationary transmission strategies, which are central in this section, and then define the Nash equilibrium of the game within this class of strategies.

   As in Section 4.1, we associate with each user i a throughput demand ρ_i (in bits per second) which it wishes to deliver over the network. The objective of each user is to minimize its average transmission power while maintaining the effective data rate at (or above) this user's throughput demand. We further assume that users always have packets to send, yet they may delay transmission to a later slot to accommodate their required throughput with minimal power investment.

   As in the previous section, our focus is on *local* and *stationary* transmission strategies, in which the transmission power decision (which includes the decision not to transmit at all) can depend only on the current state of the mobile z_i^m. The user does not have any information regarding the channel state of other users. For any given channel state, the mobile decision may include randomization over the available powers. A formal definition is provided below.

Definition 4.20   Locally stationary strategy   A *locally stationary strategy* for user i is represented by an $x_i \times (J_i + 1)$ matrix $\mathbf{q}_i$, where its (m, j) entry, denoted $q_i^{m,j}$, corresponds to the probability that user i will transmit at power Q_i^j when the observed channel state is z_i^m. As such, the set of feasible locally stationary strategies is given by

$$\Delta_i = \left\{ q_i^{m,j} \geq 0, \quad \sum_{j=0}^{J_i} q_i^{m,j} = 1 \ \forall m = 1, \ldots x_i \right\}.$$

For simplicity, we shall refer to the above defined strategy as *stationary strategy*. We also define two user-specific quantities that are derived from a given stationary strategy $\mathbf{q}_i$.

- The transmission probability $p_i(\mathbf{q}_i)$ in a slot, given by

$$p_i(\mathbf{q}_i) = \sum_{m=1}^{x_i} \pi_i^m \left(\sum_{j=1}^{J_i} q_i^{m,j} \right). \tag{4.23}$$

- The collision-free data-rate $H_i(\mathbf{q}_i)$, which stands for the average data-rate of successful transmissions, namely

$$H_i(\mathbf{q}_i) = \sum_{m=1}^{x_i} \pi_i^m \left(\sum_{j=1}^{J_i} q_i^{m,j} R_i^{m,j} \right). \tag{4.24}$$

A strategy-profile, namely a collection of (stationary) user strategies, is denoted by $\mathbf{q} \triangleq (\mathbf{q}_1, \ldots, \mathbf{q}_n)$. The notation $\mathbf{q}_{-i}$ will be used for the transmission strategies of all users but for the i-th one. Note that the probability that no user from the set $\mathcal{I} \setminus \{i\}$ transmits in a given slot is given by $\prod_{l \neq i}(1 - p_l(\mathbf{q}_l))$. Since the transmission decision of each user is independent of the decisions of other users, the expected data rate of user i, denoted $r_i(\mathbf{q})$, is given by

$$r_i(\mathbf{q}) = H_i(\mathbf{q}_i) \prod_{l \neq i}(1 - p_l(\mathbf{q}_l)). \tag{4.25}$$

A Nash equilibrium point (NE) for our model is defined below.

Definition 4.21 Nash equilibrium point   A strategy-profile $\mathbf{q} = (\mathbf{q}_1, \ldots, \mathbf{q}_i)$ is a *Nash equilibrium point* if for every $i \in \mathcal{I}$,

$$\mathbf{q}_i \in \operatorname*{argmin}_{\tilde{\mathbf{q}}_i \in \Delta_i} \left\{ \sum_{m=1}^{x_i} \pi_i^m \left(\sum_{j=1}^{J_i} \tilde{q}_i^{m,j} Q_i^j \right) : r_i(\tilde{\mathbf{q}}_i, \mathbf{q}_{-i}) \geq \rho_i \right\}. \tag{4.26}$$

Noting (4.25), it is important to emphasize that the only interaction between users is through the effective collision rate over the shared channel. This observation is significant in the context of the game *dynamics*, as the only external information that is required for best response adaptation is not user-specific and can be relatively easily measured.

4.2.2 EQUILIBRIUM ANALYSIS

In this section, we characterize the Nash equilibria (4.26) of the network under stationary transmission strategies. Our analysis starts by examining the best-response strategy of each mobile. In particular, we demonstrate that classic "water-filling" properties for optimal power control carry over to the noncooperative-game framework. We then show that one of the possible equilibrium points is best for all users in terms of their power investment.

4.2.2.1 Basic Properties

This subsection provides some basic properties of the best-response strategy. Our first result states that the throughput demands should be met with equality at every equilibrium point.

Lemma 4.22 *Let* $\mathbf{q} = (\mathbf{q}_1, \ldots, \mathbf{q}_n)$ *be a Nash equilibrium point (4.26). Then*

$$r_i(\mathbf{q}) = \rho_i. \tag{4.27}$$

Proof. The result immediately follows from (4.24)–(4.26) by noting that both the user rate and average power consumption are monotonously increasing functions of $q_i^{m,j}$, $j \in \{1, \ldots J_i\}$, $m \in \{1, \ldots x_i\}$. Indeed, if the current rate is strictly higher than required, each user may unilaterally reduce its power investment (and still hold to its requirement) by slightly decreasing some $q_i^{m,j} > 0$, $j \in \{1, \ldots J_i\}$. □

Noting (4.25), we observe that the best-response strategy can be analyzed for each user in isolation, as the overall effect of other users is manifested only through a multiplicative term which modulates the effective rate. Indeed, Lemma 4.22 indicates that the user's best response must obey $H_i(\mathbf{q}_i) = \bar{\rho}_i$, where $\bar{\rho}_i = \frac{\rho_i}{\prod_{l \neq i}(1 - p_l(\mathbf{q}_l))}$. Consequently, we shall henceforth use the latter transformation for the best-response characterization.

Consider the (discrete) rate function obtained for a given state z_i^m, and let $f_i^m(Q_i)$ be its continuous linear interpolation. Let $g_i^m(Q_i)$ be the lower concave hull of $f_i^m(Q_i)$, i.e, the smallest concave function such that $g_i^m(Q_i) \geq f_i^m(Q_i)$ (see Figure 4.4). We next assert that power levels that are not on the lower concave hull $g_i^m(Q_i)$ would not be used in any best-response strategy.

Lemma 4.23 *Consider any channel state z_i^m. Let Q_i^j be a power level whose associated rate is below $g_i^m(Q_i)$. Then $q_i^{m,j} = 0$ under every best-response strategy.*

As a consequence, of the above lemma, for any given channel state z_i^m, the interpolation of power levels that can be used (with the associated rates) creates a piecewise-linear concave function. Due to concavity, the next result immediately follows.

Lemma 4.24 *Let $\mathbf{q}_i$ be a best-response strategy. Then for every state z_i^m, there are at most two non-zero elements $q_i^{m,j}$ and $q_i^{m,k}$. Moreover, these elements correspond to adjacent power-levels Q_i^j and Q_i^k on the lower concave hull g_i^m.*

The significance of Lemma 4.24 is that the best-response for each state can be represented by a point on the concave hull graph.

Until now, we have focused on the optimal strategy within each channel state. We next provide a characterization of the best-response across the different states. To that end, we require the following definition.

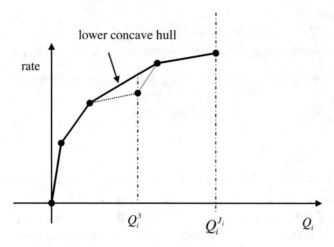

Figure 4.4: The data rate as a function of the transmission power for some fixed state z_i^m. Observe that a power level which obtains a rate which is not on the lower concave hull of the interpolation (Q_i^3 in the figure) would not be used: A convex combination of Q_i^2 and Q_i^4 can lead to a better rate at the same power cost.

Definition 4.25 rate-gain Let Q_i^j and Q_i^k ($Q_i^j < Q_i^k$) be two adjacent power-levels on the lower concave hull of g_i^m. The *rate-gain* (under state z_i^m) for these two power levels is defined as $\sigma_i^{m,k} \overset{\triangle}{=} \frac{R_i^{m,k} - R_i^{m,j}}{Q_i^k - Q_i^j}$. The next result states that higher rate-gain power levels should always be preferred. This

property is central in the characterization of the best-response, and it would have implications on the efficient calculation of the best-response, as well as on the equilibrium structure.

Lemma 4.26 *Consider two channel states z_i^m and $z_i^{m'}$. A power allocation with $q_i^{m,j} < 1$ ($j \geq 1$) and $q_i^{m',k} > 0$ ($k \geq 1$) such that $\sigma_i^{m,j} > \sigma_i^{m',k}$ is always suboptimal.*

Proof. (outline) The idea of the proof is to raise the transmission probability $q_i^{m,j}$ and decrease $q_i^{m',k}$ while preserving the same data rate. Such strategy would lower the average power investment, and the result will follow. Further details can be found in [65]. □

4.2.2.2 Equilibrium Structure

Lemma 4.26 leads to several significant properties regarding the structure of the equilibrium. As expected, better channel states would be always preferred. In addition, there always exists a best-response strategy with a single randomization of power levels. These properties are summarized in the next proposition.

Proposition 4.27   *The following properties are valid for every best-response strategy: (i) There exists some state z_i^m in which the user transmits with positive probability (i.e., $0 < \sum_{j=1}^{J_i} q_i^{m,j} \leq 1$), and further $\sum_{j=1}^{J_i} q_i^{k,j} = 1$ for $k > m$ and $\sum_{j=1}^{J_i} q_i^{k,j} = 0$ for $k < m$. (ii) There exists a best response with a single randomization; that is, for every channel state m but one, there exists some power level in $\mathcal{Q}_i$ which is used w.p. 1, i.e., $q_i^{m,j} = 1$.*

Proof. (i) Let z_i^m be the lowest channel quality at which the user transmits with positive probability. Then by construction, $\sum_{j=1}^{J_i} q_i^{k,j} = 0$ for $k < m$. For every $k > m$, let $Q_i^{j^k}$ be the smallest power level that is on the lower concave hull g_i^k. It can be easily seen that $\sigma_i^{k,j^k} > \sigma_i^{m,j}$ for every Q_j that is on the lower concave hull g_i^m. This property essentially follows from (4.21). Assume by contradiction that $\sum_{j=1}^{J_i} q_i^{k,j} < 1$; this means that $q_i^{k,j^k} < 1$. It then follows by Lemma 4.26 that such strategy cannot be optimal.

(ii) In view of Lemma 4.26, a best-response with more than a single randomization is possible if and only if two (or more) rate-gains, corresponding to active power levels of two (or more) different channel states, are the same. In that case, the same interchange argument used in the proof of Lemma 4.26 can be applied for constructing an equal power strategy that eliminates one of the randomizations. Repeating such procedure will result in a single randomization.   □

Remark 4.28   Classic results on "water-filling" optimization consider the specific rate function (4.22). For this function, it can be shown that higher powers are used for better channel states. Under our general model assumptions this is not necessarily true. Generally, an additional property of "increasing differences" is required for monotonicity in the power levels as described above. In our case, the increasing differences property holds if and only if

$$R_i^{m',j'} - R_i^{m,j'} > R_i^{m',j} - R_i^{m,j} \tag{4.28}$$

for every two indices pairs $m' > m$ and $j' > j$.

Based on Lemma 4.26, the following iterative procedure can be carried out to efficiently calculate the best-response:

1. Arrange the rate-gains in decreasing order.

2. Starting with the highest rate-gain, say $\sigma_i^{m,k} = \frac{R_i^{m,k} - R_i^{m,j}}{Q_i^k - Q_i^j}$, set $q_i^{m,k} := 1$, $q_i^{m,j} := 0$, $j \neq k$.

   Calculate the average rate via (4.24). If the total rate exceeds $\bar{\rho}_i$, find $q_i^{m,k} < 1$, $q_i^{m,j} > 0$ such that required rate is met with equality.

3. Otherwise, raise the rate by examining the next highest rate-gain and setting the associated probabilities as in Step 2. Repeat this step until obtaining the required data rate $\bar{\rho}_i$.

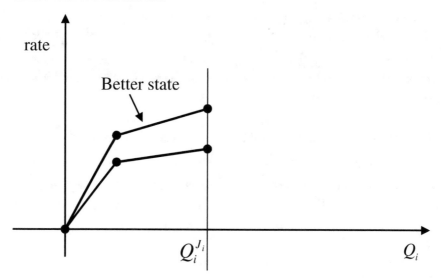

Figure 4.5: A two-state channel with two power levels available. This graph depicts the rate as a function of the used power for both states. The procedure for optimizing the power allocation starts with using the lower power level at the better state. If the obtained rate is not satisfactory, note that in this case the preceding choice is to transmit with the lower power level also at the worse state. This example demonstrates a scenario where users might transmit frequently, resulting in frequent collisions.

It immediately follows from Lemma 4.26 and Proposition 4.27 that step 3 above proceeds by either augmenting the power-level for a channel state which is at use, or initiating transmissions for a new channel state (with lower quality than the ones used so far). See Figure 4.5 for a graphical interpretation. The above procedure guarantees convergence to the best-response strategy; this property easily follows from Lemma 4.26.

4.2.2.3 Energy Efficient Equilibrium

The previous subsections established certain properties that are common to all equilibrium points. The aim of this section is to compare the possible equilibria in terms of power investment. We shall establish that the equilibrium point are *ordered* with respect to the individual users' power investment *uniformly* across all users. That is, if some user spends more power in one equilibrium than in the other, so do other users. This property immediately implies that if there exist several equilibrium points, one of them is best for all users in terms of the power investment. As before, we shall refer to this equilibrium as the *energy efficient equilibrium* (EEE).

We establish below that equilibria are ordered (component-wise) in terms of the power investment. For this result, the following lemma is required.

Lemma 4.29   *Assume that user i is a unique network user. Consider two different rate demands ρ_i and $\tilde{\rho}_i$ for that user so that $\rho_i < \tilde{\rho}_i$. Let $\mathbf{q}_i$ and $\tilde{\mathbf{q}}_i$ be best-responses for ρ_i and $\tilde{\rho}_i$, respectively. Then $p_i(\mathbf{q}_i) \leq p_i(\tilde{\mathbf{q}}_i)$.*

The above lemma immediately leads to the next theorem.

Theorem 4.30   *Let $\mathbf{q}$ and $\tilde{\mathbf{q}}$ be two equilibria so that*

$$H_i(\mathbf{q}_i) < H_i(\tilde{\mathbf{q}}_i)$$

for some user i. Then $H_l(\mathbf{q}_l) < H_l(\tilde{\mathbf{q}}_l)$ for every other user $l \neq i$. Consequently, the power investments that correspond to $\tilde{\mathbf{q}}$ are strictly higher for all users, compared to the power investments under $\mathbf{q}$.

Proof.   Recall from Lemma 4.22 that

$$r_i(\mathbf{q}) = \rho_i \tag{4.29}$$

for every equilibrium point $\mathbf{q}$. Noting (4.25), dividing the equations (4.29) of any two users i and l results in the following relation $\frac{H_i(\mathbf{q}_i)(1-p_l(\mathbf{q}_l))}{H_l(\mathbf{q}_l)(1-p_i(\mathbf{q}_i))} = \frac{\rho_i}{\rho_l}$, or

$$\frac{\left(\frac{H_i(\mathbf{q}_i)}{1-p_i(\mathbf{q}_i)}\right)}{\left(\frac{H_l(\mathbf{q}_l)}{1-p_l(\mathbf{q}_l)}\right)} = \frac{\rho_i}{\rho_l}. \tag{4.30}$$

If $H_i(\mathbf{q}_i) < H_i(\tilde{\mathbf{q}}_i)$, it follows by Lemma 4.29 that $p_i(\mathbf{q}_i) \leq p_i(\tilde{\mathbf{q}}_i)$. Hence, in order to keep the fixed ratio for $\tilde{\mathbf{q}}$, it follows that $H_l(\mathbf{q}_l) < H_l(\tilde{\mathbf{q}}_l)$. It remains to verify that the power investment at $\tilde{\mathbf{q}}$ is strictly higher compared to the equilibrium $\mathbf{q}$. Evidently, if this was not the case, users can use $\tilde{\mathbf{q}}_i$ as a starting point, by subtracting a small amount from some $\tilde{q}_i^{m,j} > 0$, $j > 1$ and obtaining a strictly lower-power allocation for $H_i(\mathbf{q}_i)$; this contradicts $\mathbf{q}$ being an equilibrium point. $\square$

The significance of Theorem 4.30 is that all mobiles are better-off at one of the equilibrium points, the EEE. The next subsection is thus dedicated to finding a simple distributed mechanism which converges to this equilibrium point.

4.2.3   BEST-RESPONSE DYNAMICS AND CONVERGENCE TO THE POWER EFFICIENT EQUILIBRIUM

As in Section 4.1.4, it turns out that the natural course of asynchronous best-response dynamics converges to the EEE. Notably, these dynamics would not require specific knowledge on other user strategies, thus they can be applicable in wireless systems.

The distributed mechanism we consider here relies on a user's best-response, which was comprehensively studied in Section 4.2.2.2 when users were not able to control their powers. Recall that in our model, the best response (BR) for user i is a solution to the following optimization problem

$$\min_{\mathbf{q}_i \in \Delta_i} \left\{ \sum_{m=1}^{x_i} \pi_i^m \left(\sum_{j=1}^{J_i} q_i^{m,j} Q_i^j \right) : H_i(\mathbf{q}_i) = \bar{\rho}_i \right\}, \qquad (4.31)$$

where $\bar{\rho}_i = \frac{\rho_i}{\prod_{l \neq i}(1 - p_l(\mathbf{q}_l))}$.

It can be shown that the BR dynamics converges to the best equilibrium under the same conditions that were required in Section 4.1.4 (see Theorem 4.18 for a precise statement of the convergence result, which is not repeated herein). Notice that the equilibrium $\mathbf{q}^b$ in our context stands for the second best equilibrium point (or a vector of ones in case of a single equilibrium point).

4.2.4   EQUILIBRIUM (IN)EFFICIENCY AND BRAESS-LIKE PARADOXES

In the previous section, we have indicated that there exists a best equilibrium in terms of power investment. Furthermore, this equilibrium can be reached through best-response dynamics. Our first objective in this section is to examine whether the best equilibrium is also socially optimal. In Section 4.1, it was shown that this is the case when a *single* power level is available to each user. We next demonstrate by means of an example that this property generally does not carry over to our current model. We then investigate the consequences of providing users with multiple power levels, and relate our observations to the Braess' paradox (see Chapter 3, Section 3.1.3).

Consider the case where a central authority, which has full information regarding the channel state distributions of every user can enforce a stationary transmission strategy (see Definition 4.20) for every user $i \in \mathcal{I}$. We consider the total power consumption as the system-wide performance criterion, namely

$$C(\mathbf{q}) \triangleq \sum_{i \in \mathcal{I}} \sum_{m=1}^{x_i} \pi_i^m \left(\sum_{j=1}^{J_i} q_i^{m,j} Q_i^j \right). \qquad (4.32)$$

Our aim is to compare the performance of the optimal centrally assigned strategy-profile to the performance at the Nash equilibrium with respect to the quantity $C(\mathbf{q})$. To that end, we use the concepts of price of anarchy (PoA) and price of stability (PoS), as in Section 4.1.2.4.

Recall that a simple example was used in the proof of Theorem 4.15, Section 4.1, to show that the PoA is generally unbounded. The example obviously holds for the current (more general) model as well. The fact that the energy efficient equilibrium can be reached by a distributed asynchronous mechanism makes the price of stability more significant, as the price of anarchy can be avoided by employing the mechanism.

We next show through a numeric example that unlike the single power case, there can be a gap between the energy efficient equilibrium and the optimal solution.

Example 4.31   We consider a wireless network of two symmetric users, with identical channel conditions and rate requirements $\rho_1 = \rho_2 = \rho$ (hence user indexes are omitted in the following). The rate per (state, power) pair is given by (4.22). The possible channel states are $\mathcal{Z} = \{0.1, 0.5, 5, 30, 80, 200\}$, and the corresponding steady-state probability vector is

$$\pi = (0.3, 0.25, 0.2, 0.12, 0.08, 0.05) \,.$$

We consider two different game instances:

- *Instance 1:* Multiple power levels are allowed for each user;

$$\mathcal{Q}_{(1)} = \{0, 1, 2, 3, 6, 8, 15\} \,.$$

- *Instance 2:* A single power level (besides zero) is allowed; $\mathcal{Q}_{(2)} = \{0, 2\}$.

For both of the above instances, Figure 4.6 depicts the per-user energy at equilibrium as a function of the required data rate ρ (note that multiple equilibria are possible for a given ρ).

The interesting region of required rates is $\rho \in [0.78, 0.92]$ (emphasized in the figure itself). For rates at this region, the EEE of Instance 2 (a single power level) obtains a lower energy investment compared to the energy efficient equilibrium of Instance 1 (multiple power levels). Noting that the single power level of 2 is one of the available powers in $\mathcal{Q}_{(1)}$, indicates that the EEE for the multiple-power case is not a system-wide optimal strategy-profile. Indeed, a simple power strategy that uses only a single power outperforms the equilibrium of Instance 1. Hence, the optimal centrally assigned strategy would obviously outperform the equilibrium policy of Instance 1 as well.

We have demonstrated through an example that the price of stability is generally larger than 1. A precise quantification for this measure obviously depends on the channel characteristics, the available powers and the assumptions on the rate functions. Explicit models and their associated price of stability are an interesting subject for future research.

A classic example for the consequences of self-interested behavior in networks is the Braess' paradox [27] (see Chapter 3, Section 3.1.3), which shows that the addition of a link in a transportation network might increase the overall traffic delay at equilibrium. We next point to Braess-like paradoxes in our network model, which concern the addition of available power levels to each user.

The first Braess-like paradox has already been demonstrated in Example 4.31. Recall that for required rates of $\rho \in [0.78, 0.92]$, Instance 2 (a single power level) outperforms Instance 1 (multiple power levels, which include the one used in Instance 2). Apparently, the addition of power levels in this example worsens user performance in terms of the average energy investment.

We next demonstrate an additional type of Braess-like paradox, which relates to the network's *capacity*. We use the term "capacity" for the (total) maximal rate that can be obtained in the network. Consider the following example.

Example 4.32   As in Example 4.31, we compare two scenarios that differ in the allowed power levels. All conditions are identical to the ones of Example 1. The only difference is that instead of Instance 2, we consider the following instance:

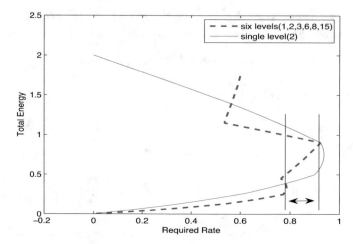

Figure 4.6: Braess-like paradox with regard to power investment. The average energy as a function of the required rate. Note that multiple equilibria are possible for a given ρ. In the marked region of required rates, the energy efficient equilibrium with a single power level outperforms the respective one with multiple power levels.

- *Instance 3:* A single power level (besides zero) is allowed; $\mathcal{Q}_{(3)} = \{0, 8\}$. Note that $\mathcal{Q}_{(3)} \subset \mathcal{Q}_{(1)}$;

Figure 4.7 presents the throughput ρ that is obtained as a function of the collision-free data rate H_i (4.24). Observe that the use of a single power level (Instance 3) increases the maximal rate ρ that can be accommodated in the network (compared to Instance 1). This example indicates that the use of multiple power levels might decrease the network capacity, due to selfish user behavior.

4.2.5   DISCUSSION

The explanation for the Braess-like paradoxes, as well as for the sub-optimality of the energy efficient equilibrium is quite intuitive, given the nature of the collision channel. In some cases, user strategies would result in frequent transmissions; this would be the case if the rate-gain at low power levels corresponding to inferior channel states is higher than the rate-gain in switching to higher power levels at good quality states. Figure 4.5 illustrates this idea. Consequently, the shared channel would be subject to frequent collisions that would lead to both unnecessary power investment and a decrease in network capacity.

It is well known that the adjustment of power levels increases capacity in single-user channels [47]. A challenging direction for future research, inspired by the above observations, is to prevent such Braess-like paradoxes in multiuser wireless networks, and even better, enlarge capacity and reduce power investment. A key role here could be given to network management that would determine the right tradeoff between user flexibility and overall performance, by assigning the power

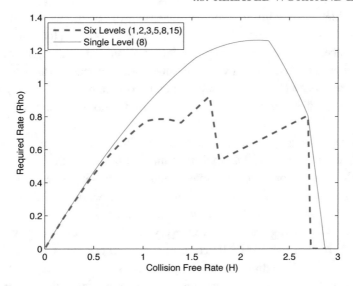

Figure 4.7: Braess-like paradox with regard to the wireless channel capacity. The obtained rate ρ as a function of the collision free rate H. Note that several equilibrium points (up to six in this particular case) are possible for each ρ. It is clearly seen that the use of a single power level may accommodate larger ρ's, compared to the case where multiple power levels are used.

levels appropriately. A central issue for future research would be how to detect suboptimal equilibria and lead the network to the best equilibrium point. A related direction is to examine the robustness of best-response mechanisms, such as the one suggested here, to changes in the transmitters population, which might occur in wireless networks. An additional research direction is to extend the reception model beyond the collision model studied in this chapter. In particular, capture models (which sometimes better represent WLAN systems) are of obvious interest. In these models, the use of higher power levels increases not only the data rate but also the chances of the individual transmission for being properly received at the base station. Hence, selfish mobiles may have a natural incentive to transmit at high power levels, which is usually not the case for the collision model studied here. It will be therefore interesting to examine whether some of the negative consequences of selfish behavior reported here will disappear, thereby reducing the gap between the energy-efficient operating point and the optimal one.

4.3   RELATED WORK AND EXTENSIONS

In this section, we briefly describe some related literature and also focus on several important model extensions.

Exploiting channel state information for increasing the network's capacity has been an on-going research topic within the information theory community (see [25] for a survey). Several papers ([7, 84] and references therein) consider uplink decentralized approaches, in which each station's transmission decision can be based on private CSI only. Nodes are assumed to operate in a cooperative manner, thus willing to accept a unified throughput-maximizing transmission policy. Papers that have considered Aloha-like random access networks from a non-cooperative perspective include [13, 44, 52, 52, 58, 99].

Numerous papers (e.g., [9, 10, 93]) have studied the equilibrium properties of power allocation games where mobile users adjust their transmission power to meet some objective (such as maximizing the throughput, or the energy investment per bit). The above papers consider a static setup where mobiles adjust their power based on average network conditions. Consequently, an equilibrium is characterized by fixed transmission powers for all users. The papers [11, 12, 56] consider the power allocation game under time-varying channel conditions, which require users to adjust their transmission power as a function of the channel state. The user objective in these references is to maximize throughput subject to power constraints. In [56], it is assumed that the channel state of a particular user is available to all others, an assumption that might be hard to justify in practice. [11, 12] consider the case where only private channel state information is available. The reception model that is studied in the above references supports multipacket-reception, as opposed to the collision channel studied in this chapter.

We list below several variations and extensions for the model considered in the previous two sections.

Capture and Multi-packet reception. An immediate extension to Sections 4.1–4.2 is to consider different reception models in the form of capture channels (i.e., a single reception is possible, even with simultaneous transmissions) or multi-packet reception. Such models are considered in [63]. It is shown in this paper that while multiple equilibrium points exist in general, one of these equilibria is uniformly best (for all users) and can be reached by best-response dynamics. A negative phenomenon, which is demonstrated in [63], is the possibility of a partial-equilibrium with starvation where stronger users (in terms of received power) satisfy their data rates, while preventing weaker ones from obtaining their respective rates.

State correlation. A recent paper [31] considers the case where the channel quality of every user is affected by global and time-varying conditions at the base station. Each user wishes to optimize its individual network utility that incorporates a natural tradeoff between throughput and power. It is shown that the equilibrium performance can be arbitrarily bad (in terms of aggregate utility); however, the efficiency loss at the best equilibrium can be bounded as a function of a technology-related parameter. Convergence to the best equilibrium is possible in special cases, however, not guaranteed in general.

Reservation and back-off mechanisms. A recent paper [67] considers the model studied in Section 4.1, with the addition of a reservation mechanism, which is used in the 802.11 standard. The medium access protocol incorporates channel reservation that relies on RTS (request-to-send) and

CTS (clear-to-send) control packets. Consequently, collisions are reduced to the relatively short periods where mobiles request channel use. The analysis reveals that the structural results that were obtained in Section 4.1 continue to hold: For feasible throughput demands, there exist exactly two Nash equilibrium points in stationary strategies, with one superior to the other uniformly over all users. Additionally, it is shown that the better equilibrium point can be obtained through best-response dynamics. However, this dynamics require individual knowledge which might no be available. Consequently, simpler myopic dynamics are suggested and shown to converge to the better equilibrium under proper initial conditions.

An additional feature of the 802.11 (and essential for its success) is the exponential back-off mechanism. It may be possible to extend the scope of the models studied in this chapter to include this important feature. As a first step in this direction, one may adopt the model suggested by Bianchi [24], which provides the means for tractable, yet fairly accurate analysis of the 802.11 performance, including the back-off mechanism.

4.4   FUTURE DIRECTIONS

To conclude this chapter, we discuss a few possible research venues within the area of wireless network games. One "persistent" direction is to keep examining the consequences of self-interested behavior in the context of new emerging technologies (e.g., WiMax, MIMO antennas). We however wish to highlight below some basic research directions, which would require the employment of additional distinctive sets of game-theoretic tools.

Battery-State Dependent Power Control. In sensor network domains, the battery of the mobile is limited and can be charged only occasionally (e.g., by solar energy). Hence, instead of considering the long-term power average (either as an objective or as a constraint), the transmitter has to be aware of its actual remaining energy. Consequently, together with varying channel conditions (as in Sections 4.1–4.2), the underlying single-user optimization task becomes a *dynamic* power control problem (rather than a static mapping from channel states to power levels). In a multiuser competitive setup, the underlying game becomes a dynamic, stochastic game, which is naturally much more involved and harder to analyze. Preliminary attempts to understand the properties of such dynamic games can be found in [61]. Recent papers [29, 30] consider the dynamic power allocation problem under arbitrarily varying channels (i.e., there is no a-priori distribution on the channel states), yet the setup is cooperative, in the sense that mobiles are interested in maximizing the total system throughput. Much remains to be explored in the noncooperative context.

Pricing, dynamics and near-potential games. Despite extensive research efforts, there is lack of predictability regarding the outcome of selfish power control in multi-packet reception domains. Since the underlying game might not have a special structure (e.g., potential or supermodular game), it becomes difficult to regulate the network to required operating points. A recent paper [32] tackles these difficulties by introducing a novel approach, termed the *potential-game approach*, which relies on approximating the underlying noncooperative game with a "close" potential game, for which prices that induce an optimal power allocation can be derived. The authors use the proximity of the original game with the approximate game to establish through Lyapunov-based analysis that natural user-update schemes (applied to the original game) converge within a neighborhood of the desired operating point, thereby inducing near-optimal performance in a *dynamical* sense. It is of interest to use the potential game approach for other wireless network games for improving the performance under selfish decision makers.

Cognitive Radios. A Cognitive Radio (CR) is a revolutionary concept that was first defined by Mitola [71] as a radio that can adapt its transmitter parameters to the environment in which it operates. It is based on the concept of Software Defined Radio (SDR) [26] that can alter parameters such as frequency band, transmission power, and modulation scheme through changes in software.

According to the Federal Communications Commission (FCC), a large portion of the assigned spectrum is used only sporadically [43]. Due to their adaptability and capability to utilize the wireless spectrum opportunistically, CRs are key enablers to efficient use of the spectrum. Under the basic model of cognitive-radio networks, Secondary Users (SUs) can use *white spaces* that are not used by the Primary Users (PUs), but they must avoid interfering with active PUs. From a game-theoretic perspective, one may view the SUs as *competing* for the available spectrum while gearing their sensing capabilities towards using the spectrum in an optimal way. These perspectives open an exciting area of research, as users actions involve not only power control and spectrum decisions, but also sensing decisions (which spectrum channels to sense and when). The latter set of decisions is closely related to the theoretical frameworks of experts and multi-armed bandits. Indeed, some recent papers (e.g., [14] and references therein) study the consequences of distributed sensing on network performance by using these frameworks for algorithm design and analysis.

CHAPTER 5

Future Perspectives

We conclude this monograph by outlining some high-level directions for future research in the field of network games. As this field is interdisciplinary by nature, we believe that research should consist of two commensurate paths:

1. Develop new game-theoretic *models* to represent interactions among heterogenous users in current and future communication networks. This also involves identifying the right equilibrium notion that captures the strategic interactions among the players.

2. Develop novel game-theoretic *tools* for the analysis of complex user interaction in noncooperative domains. The objective here is to employ such tools in the design of new network architectures and protocols.

Game theory provides numerous solution concepts for the understanding of noncooperative user interaction. However, some of the notions that have been developed are not yet fully deployed in networked systems. These include:

Correlated equilibrium. To the best of our knowledge, there are no convincing applications in networked systems that exploit the advantages of the correlated equilibrium concept, despite their appealing static and dynamic properties. It is, therefore, of great interest to identify potential networking applications in which correlation between users can be used to improve system efficiency.

Dynamic (Markov) games. As indicated in Chapter 4, static one-shot games fail short to model certain user interactions in networks. As exemplified therein in the context of sensor networks the actual battery condition of a mobile sensor could be the most important factor in its current behavior and future actions. This gives rise to a dynamic game with an evolving state. Another example is a peer-to-peer network, in which the decision of whether to download a certain content from a central server determines if and how the peers would be able to benefit from that content. An important research direction is therefore to employ the theory of dynamic games in the relevant network application.

In our opinion, the need for new game-theoretic tools arises from two related reasons. First, many games over networks do not have a special structure (e.g., potential or supermodular), hence these games are often hard to analyze. Second, little can be said about the convergence properties of user dynamics, unless, again, the underlying game has special structure. Motivated by these issues, we suggest the following research directions:

Analyzing games by their relation to games with structure. As mention earlier, we have recently introduced a novel *potential game approach* and applied it in the context of distributed power allocation in multi-cell wireless networks [32]. This approach relies on approximating the underlying noncooperative game with a "close" potential game, for which simple incentive schemes that induce a system-optimal power allocation can be derived. We believe that this approach can be applied to other resource allocation problems, such as power control in multi-channel systems, and routing in wireline and wireless networks.

An important direction is to extend the above idea beyond potential games. As we elaborated in this monograph, there are additional classes of games with predictable dynamics outcomes. These include ordinal potential games and supermodular games. By extending the potential game approach to other games with special structure, the hope is to be able to cover numerous networking applications, thereby improving the analysis and controllability thereof.

Dynamic efficiency. In relation to the above item, the ability to enhance the predictability of the dynamics in noncooperative networks motivates the study of the *efficiency* of such system. The bulk of the research that has dealt with quantifying the efficiency (or efficiency loss) in networks focused on *static* measures. The well-studied notions of PoA and PoS compare the (Nash) equilibrium performance to the socially optimal operating point. However, these measures do not capture the underlying dynamics of the system. For example, some of the equilibria might not be reached by natural user dynamics, thereby diminishing the relevance of the static efficiency loss measures. It is therefore of great interest to define and study measures of efficiency loss in the *dynamical* sense; e.g., comparing the possible limiting outcomes of best-response dynamics to the socially-optimal operating point. While this and other related topics could have been considered under very specialized game structures, the potential game approach and similar approaches may allow one to make progress for general, unstructured, non-cooperative games.

An additional topic for future work relates to the underlying assumptions on utility functions in network games. As indicated in [33], the commonly-adopted assumption of selfishness has been repeatedly questioned by economists and psychologists. Experiments have shown that in some network environments, the users do not necessarily act selfishly to optimize their own performance, and their behavior can be regarded as either *altruistic* or *malicious*. Accordingly, an active research area has been to incorporate altruistic or malicious characteristics into the utility functions, and examine the corresponding equilibrium properties; see, e.g., [33].

As a closing comment, we mention that while the emphasis in this monograph has been on communication networks, game theory has become a dominant tool for the analysis and design of emerging applications such as sponsored search auctions and social networks. Indeed, the self-interested behavior of users plays a central role in these domains. Therefore, we strongly believe that the research in the field of networking games will continue playing an important role in the better understanding and design of present and future network systems.

Bibliography

[1] *IEEE 802.11 standards*, Available from `http://standards.ieee.org/getieee802/802.11.html`. 103

[2] D. Acemoglu, K. Bimpikis, and A. Ozdaglar, *Price and capacity competition*, Games and Economic Behavior **66** (2009), no. 1, 1–26. DOI: 10.1016/j.geb.2008.06.004 96

[3] D. Acemoglu, R. Johari, and A. Ozdaglar, *Partially optimal routing*, Journal on Selected Areas in Communications **25** (2007), no. 6, 1148–1160. DOI: 10.1109/JSAC.2007.070809 79, 87

[4] D. Acemoglu and A. Ozdaglar, *Flow control, routing, and performance from service provider viewpoint*, LIDS report, WP-1696, 2004. 19, 88

[5] ———, *Competition and efficiency in congested markets*, Mathematics of Operations Research **32** (2007), no. 1, 1–31. DOI: 10.1287/moor.1060.0231 19, 20, 88, 90, 91, 92, 93, 94

[6] ———, *Competition in parallel-serial networks*, Journal on Selected Areas in Communications **25** (2007), no. 6, 1180–1192. DOI: 10.1109/JSAC.2007.070812 19, 88, 92, 94, 95

[7] S. Adireddy and L. Tong, *Exploiting decentralized channel state information for random access*, IEEE Transactions on Information Theory **51** (2005), no. 2, 537–561. DOI: 10.1109/TIT.2004.840878 130

[8] R.K. Ahuja, T.L. Magnanti, and J.B. Orlin, *Network flows*, Prentice-Hall, Englewood Cliffs, New Jersey, 1993. xiii

[9] T. Alpcan, T. Başar, R. Srikant, and E. Altman, *CDMA uplink power control as a noncooperative game*, Wireless Networks **8** (2002), 659–670. DOI: 10.1023/A:1020375225649 130

[10] E. Altman and Z. Altman, *S-modular games and power control in wireless networks*, IEEE Transactions on Automatic Control **48** (2003), no. 5, 839–842. DOI: 10.1109/TAC.2003.811264 58, 130

[11] E. Altman, K. Avrachenkov, I. Menache, G. Miller, B. Prabhu, and A. Shwartz, *The water-filling game in fading multiple access channel*, IEEE Transactions on Automatic Control **54** (2009), no. 10, 2328–2340. DOI: 10.1109/TIT.2008.920340 130

[12] E. Altman, K. Avrachenkov, G. Miller, and B. Prabhu, *Discrete power control: Cooperative and non-cooperative optimization*, INFOCOM, 2007, pp. 37–45. DOI: 10.1109/INFCOM.2007.13 130

[13] E. Altman, R. El-Azouzi, and T. Jimenez, *Slotted Aloha as a game with partial information*, Computer Networks **45** (2004), no. 6, 701–713. DOI: 10.1016/j.comnet.2004.02.013 130

[14] A. Anandkumar, N. Michael, and A.K. Tang, *Opportunistic Spectrum Access with Multiple Users: Learning under Competition*, INFOCOM, 2010. DOI: 10.1109/INFCOM.2010.5462144 132

[15] T. Basar and G.J. Olsder, *Dynamic noncooperative game theory*, Academic Press, London/New York, 1989. 4

[16] T. Basar and R. Srikant, *Revenue-maximizing pricing and capacity expansion in a many-users regime*, Proc. of INFOCOM, 2002. DOI: 10.1109/INFCOM.2002.1019271 88

[17] _____ , *A stackelberg network game with a large number of followers*, Journal of Optimization Theory and Applications **115** (2002), no. 3, 479–490. DOI: 10.1023/A:1021294828483 88

[18] M. Beckmann, C.B McGuire, and C. B. Winsten, *Studies in the economics of transportation*, Yale University Press, 1956. 74

[19] D. Bertsekas and R. Gallager, *Data networks*, Prentice-Hall, 1992. 114

[20] D.P. Bertsekas and R.G. Gallager, *Data networks*, Prentice-Hall, Englewood Cliffs, New Jersey, 1992. xiii

[21] D.P. Bertsekas, A. Nedić, and A.E. Ozdaglar, *Convex analysis and optimization*, Athena Scientific, Cambridge, Massachusetts, 2003. 22, 87

[22] D.P. Bertsekas and J.N. Tsitsiklis, *Parallel and distributed computation*, Prentice-Hall, Englewood Cliffs, New Jersey, 1989. xiii

[23] U. Bhaskar, L. Fleischer, D. Hoy, and C. C. Huang, *Equilibria of atomic flow games are not unique*, SODA, SIAM, 2009, pp. 748–757. DOI: 10.1023/A:1021294828483 97

[24] G. Bianchi, *Performance analysis of the IEEE 802.11 distributed coordination function*, IEEE Journal on Selected Areas in Communications **18** (2000), no. 3, 535–547. DOI: 10.1109/49.840210 131

[25] E. Biglieri, J. Proakis, and S. Shamai, *Fading channels: Information-theoretic and communications aspects*, IEEE Transactions on Information Theory **44** (1998), no. 6, 2619–2692. DOI: 10.1109/18.720551 99, 102, 104, 130

[26] Vanu G. Bose, Alok B. Shah, and Michael Ismert, *Software Radios for Wireless Networking*, Proc. IEEE INFOCOM'98, Apr. 1998. 132

[27] D. Braess, *Uber ein paradoxon aus der verkehrsplanung*, Unternehmensforschung **12** (1969), 258 – 268. 19, 71, 77, 127

[28] G. Brown, *Iterative solution of games by fictitious play*, Activity Analysis of Production and Allocation (T. Koopmans, ed.), John Wiley and Sons, New York, New York, 1951. 45

[29] N. Buchbinder, L. Lewin-Eytan, I. Menache, J. Naor, and A. Orda, *Dynamic power allocation under arbitrary varying channels - an online approach*, INFOCOM, 2009, pp. 145–153. DOI: 10.1109/INFCOM.2009.5061916 131

[30] _____, *Dynamic power allocation under arbitrary varying channels - the multi-user case*, INFOCOM, 2010. DOI: 10.1109/INFCOM.2010.5462067 131

[31] O. Candogan, I. Menache, A. Ozdaglar, and P. A. Parrilo, *Competitive scheduling in wireless collision channels with correlated channel state*, GameNets, 2009. DOI: 10.1109/GAMENETS.2009.5137452 130

[32] _____, *Near-optimal power control in wireless networks - a potential game approach*, INFOCOM, 2010. DOI: 10.1109/INFCOM.2010.5462017 131, 133

[33] P. Chen and D. Kempe, *Altruism, selfishness, and spite in traffic routing*, Proc. of Electronic Commerce (EC), 2008. DOI: 10.1145/1386790.1386816 134

[34] R. Cominetti, J. R. Correa, and N. E. Stier-Moses, *The impact of oligopolistic competition in networks*, Operations Research **57** (2009), no. 6, 1421–1437. DOI: 10.1287/opre.1080.0653 97

[35] J. R. Correa, A. S. Schulz, and N. E. Stier-Moses, *Selfish routing in capacitated networks*, Mathematics of Operations Research **29** (2004), no. 4, 961–976. DOI: 10.1287/moor.1040.0098 84

[36] _____, *A geometric approach to the price of anarchy in nonatomic congestion games*, Games and Economic Behavior **64** (2008), 457–469. DOI: 10.1016/j.geb.2008.01.001 84, 90

[37] J. R. Correa and N. E. Stier-Moses, *Wardrop equilibria*, Wiley Encyclopedia of Operations Research and Management Science (J. J. Cochran, ed.), John Wiley and Sons, 2010, Forthcoming. 20

[38] C. Courcoubetis and R. Weber, *Pricing communication networks: Economics, technology and modelling*, Wiley, 2003. 96

[39] S. C. Dafermos and F. T. Sparrow, *The traffic assignment problem for a general network*, J. of Research of the National Bureau of Standards **73** (1969), no. 2, 91–118. 85

[40] P. Dasgupta and E. Maskin, *The existence of equilibrium in discontinuous economic games. 1: Theory*, Review of Economic Studies **53** (1986), 1–26. 29

[41] _____, *The existence of equilibrium in discontinuous economic games. 2: Applications*, Review of Economic Studies **53** (1986), 27–42. 29

[42] L. A. DaSilva, *Pricing for qos-enabled networks: a survey*, IEEE Communication Surveys and Tutorials **3** (2000), no. 2, 2–8. DOI: 10.1109/COMST.2000.5340797 96

[43] FCC, *ET Docket No. 03-222, Notice of Proposed Rule Making and Order*, Dec. 2003. 132

[44] A. Fiat, Y. Mansour, and U. Nadav, *Efficient contention resolution for selfish agents*, Symposium on Discrete Computing (SODA), 2007, pp. 179 – 188. 130

[45] D. Fudenberg and D.K. Levine, *The theory of learning in games*, MIT Press, Cambridge, Massachusetts, 1999. 45, 49, 52

[46] D. Fudenberg and J. Tirole, *Game theory*, MIT Press, Cambridge, Massachusetts, 1991. 43, 56, 62, 114

[47] A. J. Goldsmith and P. P. Varaiya, *Capacity of fading channels with channel side information*, IEEE Transactions on Information Theory **43** (1997), no. 6, 1986–1992. DOI: 10.1109/18.641562 128

[48] M. Haviv and R. Hassin, *To queue or not to queue: Equilibrium behavior in queueing systems*, 2003. 97

[49] A. Hayrapetyan, E. Tardos, and T. Wexler, *A network pricing game for selfish traffic*, Proc.of ACM SIGACT-SIGOPS Symposium on Principles of Distributed Computing, 2005. DOI: 10.1145/1073814.1073869 94

[50] J. Huang, R. Berry, and M. L. Honig, *Distributed interference compensation for wireless networks*, IEEE Journal on Selected Areas in Communications **24** (2006), no. 5, 1074–1085. DOI: 10.1109/JSAC.2006.872889 58

[51] X. Huang, A. Ozdaglar, and D. Acemoglu, *Efficiency and braess' paradox under pricing in general networks*, IEEE Journal on Selected Areas in Communication **24** (2006), no. 5, 977–991. DOI: 10.1109/JSAC.2006.872879 91

[52] Y. Jin and G. Kesidis, *Equilibiria of a noncooperative game for heterogeneous users of an ALOHA network*, IEEE Comm. Letters **6** (2002), no. 7, 282–284. DOI: 10.1109/LCOMM.2002.801326 130

[53] S. Kakutani, *A generalization of brouwerŌs fixed point theorem*, Duke Mathematical Journal **8** (1941), no. 3, 457–459. DOI: 10.1215/S0012-7094-41-00838-4 22

[54] F.P. Kelly, A.K. Maulloo, and D.K. Tan, *Rate control for communication networks: Shadow prices, proportional fairness, and stability*, Journal of the Operational Research Society **49** (1998), 237–252. 58

[55] E. Koutsoupias and C. Papadimitriou, *Worst-case equilibria*, Proc. of the 16th Annual Symposium on Theoretical Aspects of Computer Science, 1999, pp. 404–413. DOI: 10.1007/3-540-49116-3_38 75

[56] L. Lai and H. El-Gamal, *The water-filling game in fading multiple access channel*, IEEE Transactions on Information Theory (2006), Submitted. DOI: 10.1109/TIT.2008.920340 130

[57] S. Low and D.E. Lapsley, *Optimization flow control, I: Basic algorithm and convergence*, IEEE/ACM Transactions on Networking **7** (1999), no. 6, 861–874. DOI: 10.1109/90.811451 58

[58] B. MacKenzie and S. B. Wicker, *Stability of slotted aloha with multipacket reception and selfish users*, Proceedings of INFOCOM, 2003, pp. 1583–1590. DOI: 10.1109/INFCOM.2003.1209181 130

[59] N. B. Mandayam, S. B. Wicker, J. Walrand, T. Basar, J. Huang, and D. P. Palomar (editors), *Special issue on game theory in communication systems*, Journal of Selected Areas in Communications **26** (2008), no. 7. 99

[60] A. Mas-Colell, M.D. Whinston, and J.R. Green, *Microeconomic theory*, Oxford University Press, 1995. 3

[61] I. Menache and E. Altman, *Battery-state dependent power control as a dynamic game*, WiOpt, 2008. DOI: 10.1109/WIOPT.2008.4586072 131

[62] I. Menache and N. Shimkin, *Capacity management and equilibrium for proportional qos*, IEEE/ACM Transactions on Networking **16** (2008), no. 5, 1025–1037. DOI: 10.1109/TNET.2007.911430 96

[63] _____, *Decentralized rate regulation in random access channels*, INFOCOM, 2008, pp. 394–402. DOI: 10.1109/INFOCOM.2008.86 130

[64] _____, *Efficient rate-constrained nash equilibrium in collision channels with state information*, INFOCOM, 2008, pp. 403–411. DOI: 10.1109/INFOCOM.2008.87 102, 111

[65] _____, *Noncooperative power control and transmission scheduling in wireless collision channels*, SIGMETRICS, 2008, pp. 349–358. DOI: 10.1145/1384529.1375497 102, 122

[66] _____, *Rate-based equilibria in collision channels with fading*, IEEE Journal on Selected Areas in Communications **26** (2008), no. 7, 1070–1077. DOI: 10.1109/JSAC.2008.080905 102, 111, 113, 114, 116

[67] _____, *Reservation-based distributed medium access in wireless collision channels*, Telecommunications Systems (2010), Available online. DOI: 10.4108/ICST.VALUETOOLS2008.4475 102, 115, 130

[68] F. Meshkati, H. V. Poor, and S. C. Schwartz, *Energy-efficient resource allocation in wireless networks*, IEEE Signal Processing Magazine **24** (2007), no. 3, 58–68. DOI: 10.1109/MSP.2007.361602 101

[69] P. Milgrom and J. Roberts, *Rationalizability, learning and equilibrium in games with strategic complementarities*, Econometrica **58** (1990), 1255–1278. DOI: 10.2307/2938316 56, 57

[70] P. Milgrom and C. Shannon, *Monotone comparative statics*, Econometrica **62** (1994), 157–180. 56

[71] J. Mitola III, *Cognitive Radio for Flexible Mobile Multimedia Communications*, ACM/Kluwer MONET **6** (2001), no. 5, 435–441. DOI: 10.1109/MOMUC.1999.819467 132

[72] K. Miyasawa, *On the convergence of learning processes in a 2x2 nonzero sum game*, Research Memo 33, Princeton, 1961. 49

[73] A. F. Molisch, *Wireless communications*, John Wiley and Sons, 2005. 101

[74] D. Monderer, D. Samet, and A. Sela, *Belief affirming in learning processes*, Mimeo, Technion, 1994. 50

[75] D. Monderer and L.S. Shapley, *Fictitious play property for games with identical interests*, Journal of Economic Theory **68** (1996), 258–265. DOI: 10.1006/jeth.1996.0014 49

[76] _____, *Potential games*, Games and Economic Behavior **14** (1996), 124–143. DOI: 10.1006/game.1996.0044 63

[77] R. Myerson, *Game theory: Analysis of conflict*, Harvard University Press, Cambridge, Massachusetts, 1991. 26

[78] J. Nash, *Non-cooperative games*, Annals of Mathematics **54** (1951), no. 2, 286–295. 23

[79] P. Njoroge, A. Ozdaglar, N. Stier-Moses, and G. Weintraub, *Competition, market coverage, and quality choice in interconnected platforms*, in Proc. of NetEcon, 2009, Available online: http://netecon.seas.harvard.edu/NetEcon09/index09.html. 97

[80] A. M. Odlyzko, *Paris metro pricing for the internet*, Proc. of the first ACM conference on Electronic Commerce (EC), 1999, pp. 140 – 147. DOI: 10.1145/336992.337030 96

[81] A. Orda, R. Rom, and N. Shimkin, *Competitive routing in multi-user communication networks*, IEEE/ACM Trans. on Networking **1** (1993), no. 3, 510–521. DOI: 10.1145/336992.337030 97

[82] A. Ozdaglar, *Price competition with elastic traffic*, Networks (2008), Published online. DOI: 10.1002/net.v52:3 94

[83] A. Ozdaglar and R. Srikant, *Incentives and pricing in communication networks*, Algorithmic Game Theory (N. Nisan, T. Roughgarden, E. Tardos, and V. Vazirani, eds.), Cambridge University Press, 2007. 19

[84] X. Qin and R. Berry, *Distributed approaches for exploiting multiuser diversity in wireless networks*, IEEE Transactions on information theory **52** (2006), no. 2, 392–413. DOI: 10.1109/TIT.2005.862103 104, 130

[85] J. Rexford, *Route optimization in ip networks*, in Handbook of Optimization in Telecommunications, Springer Science + Business, Kluwer Academic Publishers, 2006. 79

[86] O. Richman and N. Shimkin, *Topological uniqueness of the nash equilibrium for atomic selfish routing*, Mathematics of Operations Research **32** (2007), no. 1, 215–232. DOI: 10.1287/moor.1060.0229 97

[87] J. Robinson, *An iterative method of solving a game*, Annals of Mathematical Statistics **54** (1951), 296–301. 49

[88] J.B. Rosen, *Existence and uniqueness of equilibrium points for concave n-person games*, Econometrica **33** (1965), no. 3. 32

[89] R.W. Rosenthal, *A class of games possessing pure-strategy nash equilibria*, International Journal of Game Theory **2** (1973), 65–67. DOI: 10.1007/BF01737559 66

[90] T. Roughgarden, *Selfish routing and the price of anarchy*, MIT Press, 2005. 77, 78, 97

[91] T. Roughgarden and E. Tardos, *How bad is selfish routing*, Journal of the ACM **49** (2002), 236–259. DOI: 10.1145/506147.506153 76

[92] E. Sabir, R. El-Azouzi, V. Kavitha, Y. Hayel, and E. Bouyakhf, *Stochastic learning solution for constrained nash equilibrium throughput in non saturated wireless collision channels*, Proc. of VALUETOOLS, 2009, pp. 1–10. DOI: 10.4108/ICST.VALUETOOLS2009.7764 116

[93] C. U. Saraydar, N. B. Mandayam, and D. J. Goodman, *Efficient power control via pricing in wireless data networks*, IEEE Transactions on Communications **50** (2002), no. 2, 291–303. DOI: 10.1109/26.983324 58, 130

[94] L.S. Shapley, *Some topics in two-person games*, Advances in Game Theory (M. Drescher, L.S. Shapley, and A.W. Tucker, eds.), Princeton University Press, Princeton, 1964. 50

[95] A. Simsek, A. Ozdaglar, and D. Acemoglu, *Uniqueness of generalized equilibrium for box constrained problems and applications*, Proc. of Allerton Conference, 2005. 35

[96] _____, *A generalized Poincare-Hopf theorem for compact nonsmooth regions*, Mathematics of Operations Research **32** (2007), no. 1, 193–214. DOI: 10.1287/moor.1060.0235 35

[97] _____, *Local indices for degenerate variational inequalities*, Mathematics of Operations Research **33** (2008), no. 2, 291–301. DOI: 10.1287/moor.1070.0299 35

[98] N. Stein, P.A. Parrilo, and A. Ozdaglar, *Correlated equilibria in continuous games: Characterization and computation*, to appear in Games and Economic Behavior (2010). DOI: 10.1109/CDC.2007.4434890 19

[99] J. Sun and E. Modiano, *Opportunistic power allocation for fading channels with non-cooperative users and random access*, IEEE BroadNets Wireless Networking Symposium, 2005. DOI: 10.1109/ICBN.2005.1589639 130

[100] D. Topkis, *Supermodularity and complementarity*, Princeton University Press, Princeton, 1998. 56

[101] D.M. Topkis, *Equilibrium points in nonzero-sum n-person submodular games*, SIAM Journal of Control and Optimization **17** (1979), no. 6, 773—-787. DOI: 10.1137/0317054 56, 57, 60

[102] J.N. Tsitsiklis and D.P. Bertsekas, *Distributed asynchronous optimal routing in data networkss*, IEEE Transactions on Automatic Control **31** (1986), no. 4, 325–332. DOI: 10.1109/TAC.1986.1104261 xiii

[103] X. Vives, *Nash equilibrium with strategic complementarities*, Journal of Mathematical Economics **19** (1990), no. 3, 305–321. DOI: 10.1016/0304-4068(90)90005-T 56

[104] _____, *Oligopoly pricing*, MIT Press, Cambridge, Massachusetts, 2001. 56

[105] J. G. Wardrop, *Some theoretical aspects of road traffic research*, Proc. of the Institute of Civil Engineers, Part II **1** (1952), 325–378. 72

[106] M. Zorzi, *Mobile radio slotted ALOHA with capture and diversity*, INFOCOM, 1995, pp. 121–128. DOI: 10.1007/BF01202545 101

Authors' Biographies

ISHAI MENACHE

Ishai Menache received his PhD degree in Electrical Engineering from the Technion, Israel Institute of Technology, in 2008. Prior to his graduate studies, he worked for a couple of years in Intel, as an engineer in the networks communication group. Until recently, he was a postdoctoral associate at the Laboratory for Information and Decision Systems in MIT. He is currently a visiting researcher at Microsoft Research New England, focusing on pricing and resource allocation aspects of Cloud Computing. Dr. Menache's broader areas of interest include communication networks, game theory and machine learning. He is a recipient of the Marie Curie Outgoing International Fellowship.

ASUMAN OZDAGLAR

Asuman Ozdaglar received the B.S. degree in electrical engineering from the Middle East Technical University, Ankara, Turkey, in 1996, and the S.M. and the Ph.D. degrees in electrical engineering and computer science from the Massachusetts Institute of Technology, Cambridge, in 1998 and 2003, respectively.

Since 2003, she has been a member of the faculty of the Electrical Engineering and Computer Science Department at the Massachusetts Institute of Technology, where she is currently the Class of 1943 Associate Professor. She is also a member of the Laboratory for Information and Decision Systems and the Operations Research Center. Her research interests include optimization theory, with emphasis on nonlinear programming and convex analysis, game theory, with applications in communication, social, and economic networks, and distributed optimization and control. She is the co-author of the book entitled "Convex Analysis and Optimization" (Athena Scientific, 2003).

Professor Ozdaglar is the recipient of a Microsoft fellowship, the MIT Graduate Student Council Teaching award, the NSF Career award, and the 2008 Donald P. Eckman award of the American Automatic Control Council. She served on the Board of Governors of the Control System Society in 2010. She is currently the chair of the working group "Game-Theoretic Methods in Networks" under the Technical Committee "Networks and Communications Systems" of the IEEE Control Systems Society and serves as an associate editor for the area Optimization Theory, Algorithms and Applications for the Asia-Pacific Journal of Operational Research.

Printed in the United States
by Baker & Taylor Publisher Services